Balcony Solar for Renters

The Complete Guide to Plug-and-Play Solar Power for Apartments

Transform Your Apartment into a Power-Generating Space—No Roof Required

Maurice Heath Owen

Balcony Solar for Renters: The Complete Guide to Plug-and-Play Solar Power for Apartments

ISBN: 978-1-923646-00-1

First Edition: November 2025

Renewable Ink Publishing

DISCLAIMER

This book provides general information for educational purposes only. Solar technology, regulations, and products change rapidly. Information is current as of publication but may have changed.

All personal names, individual scenarios, and case examples used throughout this book (including but not limited to Sarah, Kevin, Marcus, Maria, Tom, Rachel, David, Jason, Jennifer, Melissa, and others) are fictional composite and created for illustrative purposes only. Any resemblance to actual persons, living or dead, or actual events is purely coincidental. These examples are composite scenarios based on typical situations and do not represent specific individuals or their experiences.

Building codes, electrical regulations, and solar requirements vary by jurisdiction and change frequently. Readers must verify current regulations in their specific location before installation. Always check with local authorities, utilities, and property management.

This book does not constitute electrical, legal, financial, or safety advice. The author is not a licensed electrician, attorney, financial advisor, or certified installer. Consult qualified professionals for specific guidance.

Product and company names (including EcoFlow, Jackery, Anker, MANLY Battery, BigBlue, Bright Saver, Robinsun, Enphase, Hoymiles, Renogy, and others) are for informational purposes only. No endorsement

Table of Contents

About The Author **1**

Chapter 1: The Balcony Solar Revolution **3**

The 53 Million Americans Left Behind 3

How Germany Accidentally Started a Revolution 4

Utah Becomes America's Unlikely Solar Pioneer 5

Why 2025 Is the Tipping Point 6

Real People, Real Results 9

Breaking the Homeowner Monopoly 10

The Portable Power Advantage 10

Chapter 2: Understanding Plug-and-Play Solar Technology **15**

What Is Balcony Solar, Really? 15

How Plug-In Solar Actually Works 16

Power Generation Reality Check 18

The 10-30% Offset Sweet Spot 19

Microinverters vs. String Inverters 20

Safety Features That Matter 21

Why These Systems Can't Power Your Entire Apartment 22

Comparing Your Options 23

What You Need to Remember 24

Chapter 3: State-by-State Regulatory Navigator **26**

The Utah Model: How One State Got It Right 26

States to Watch: The Momentum Is Building 28

California's Unique Approach 29

Understanding Interconnection Agreements 30

National Electrical Code: What Exists Now 32
UL Certification: Why It Matters 33
Local Utility Policies: The Wild West 34
HOA and Condo Association Regulations 35
Building Code Variations by Metro Area 36
Navigating the Gray Area Safely 37
What Matters Right Now 38
Chapter 4: The Landlord Negotiation Playbook 40
When You Actually Need Permission 40
Portable vs. Permanent Installations 42
The "No Structural Modifications" Argument 43
Building Your Business Case for Landlords 44
Permission Request Templates That Work 45
What Your Lease Actually Says 47
Tenant Rights You Should Know 48
Germany's Tenant Privilege Approach 49
Insurance and Liability Considerations 50
Documentation That Protects You 51
When Your Landlord Says No 52
Organizing with Other Tenants 53
What Actually Happens Most of the Time 54
Action Plan 54
Chapter 5: Urban Installation Mastery 57
Why Direction Matters 57
Making Peace with Suboptimal Orientation 58
Shading Analysis Tools You Can Use 59
Working with Partial Shade 61

Balcony Types and Mounting Solutions 62
Maximizing Limited Square Footage 64
High-Rise Challenges Nobody Talks About 65
Fire Escape Regulations 66
Mounting Systems Compared 67
DIY Installation Step-by-Step 69
When to Hire an Electrician 73
Seasonal Optimization Strategies 74
Keeping Neighbors and Management Happy 74
Your Installation Gameplan 75
Chapter 6: Battery Storage for Apartments 77
The Evening Use Problem 77
Battery Sizing That Makes Sense 78
Lithium-ion vs LiFePO4 Battery Chemistry 79
Modular Battery Systems 81
Smart Battery Management 82
Self-Consumption Modes 83
Backup Power Reality Check 84
Indoor vs Outdoor Battery Placement 85
Temperature Considerations 87
Product Recommendations 88
ROI Analysis for Battery Storage 90
Safety Certifications That Matter 92
What Works for Apartments 93
Chapter 7: Financial Analysis and ROI Calculators 95
High Electricity Rate Cities 96
Medium Electricity Rate Markets 97

Lower Electricity Rate Areas 99
Climate and Sun Hours 100
Breaking Down System Costs..101
DIY vs Professional Installation 103
Federal Tax Credit Details 105
State and Local Incentives 106
Utility Rate Structures..107
Realistic Monthly Savings 109
The Portability Value Equation..110
Financing Options 111
Resale Value Considerations..112
Environmental ROI 114
Your Numbers, Your Decision 115
Chapter 8: Product Selection and Installation Guide..............116
Entry-Level Systems 116
Mid-Range Systems 118
Premium Systems...120
Trusted Manufacturers 122
US Market Availability (2025 Update)...................................123
Reading Solar Panel Datasheets...125
Inverter Selection 126
Monitoring Systems 128
Installation Tools You Actually Need.....................................129
Safety Testing Before First Use 130
Troubleshooting Common Issues...131
Maintenance Requirements..133
When and How to Upgrade..134

Moving with Your System 134
Warranty Considerations 135
What You Need to Do Next 137
Appendix A: State-by-State Regulatory Quick Reference 138
Appendix B: Landlord Permission Letter Template 140
Appendix C: ROI Calculator Spreadsheet 143
Appendix D: Installation Safety Checklist 145
Appendix E: Product Comparison Matrix 147
Appendix F: Utility Contact Information by State 149
Appendix G: Glossary of Technical Terms 151
Appendix H: Additional Resources and Community Groups 153
References 157

About The Author

Maurice Heath Owen is a renewable energy advocate and researcher who spent three years investigating why 102 million American renters remain locked out of the solar revolution while European apartment dwellers have been generating their own electricity for over a decade.

His interest in balcony solar began when he moved into a fifth-floor apartment in 2021 and discovered that his $180 monthly electric bills gave him zero options for reducing costs. Traditional rooftop solar wasn't available. Community solar had long waitlists. Energy efficiency measures barely made a dent. Meanwhile, he watched news coverage of Germany installing hundreds of thousands of plug-and-play balcony solar systems in apartments—technology that simply didn't exist in the US market.

Frustrated by the information gap, Maurice began researching European balcony solar systems, US regulatory barriers, and the technical requirements for safe apartment installation. What started as personal research evolved into a mission to document everything American renters need to know about this emerging technology.

Over three years, Maurice interviewed apartment dwellers who successfully installed balcony solar, consulted with solar engineers about urban installation challenges, researched state-by-state regulations, analyzed utility policies across the country, and tested equipment to understand what works in real-world apartment conditions. He tracked legislative developments as Utah became the first state to explicitly legalize plug-and-play solar in 2025 and watched as other states began following suit.

This book represents the culmination of that research—the comprehensive US guide that didn't exist when Maurice started his journey. It's written for renters, by someone who understands the unique challenges renters face: limited budgets, uncertain tenure, landlord restrictions, and the need for portable solutions that move from apartment to apartment.

Maurice lives in Denver, Colorado, where his south-facing balcony generates approximately 2.1 kWh of solar power daily using a 600-watt system he installed himself. When he's not monitoring his solar production or researching renewable energy policy, he advocates for expanded solar access for renters and works to help apartment communities adopt balcony solar programs.

He believes the solar revolution shouldn't be limited to homeowners. The 102 million Americans who rent deserve access to clean, affordable, self-generated electricity.

This book is his contribution to making that happen.

Chapter 1: The Balcony Solar Revolution

Jessica Brookes stood on her tiny Brooklyn balcony, phone in hand, staring at her latest electric bill. One hundred seventy-three dollars. For a 650-square-foot apartment. She'd already switched to LED bulbs, unplugged everything at night, and kept her thermostat at levels her friends called "arctic" in summer and "polar" in winter. Nothing helped.

"There's got to be something else I can do," she muttered, scrolling through energy-saving tips she'd already tried. Then an ad popped up. A solar panel. On a balcony. In Germany. She clicked.

Two hours later, Jessica discovered something that would change not just her electric bill, but her entire relationship with energy. Turns out, millions of apartment dwellers across Europe were generating their own solar power from balconies no bigger than hers. No roof required. No landlord permission needed. Just plug it in and start saving.

But here's the thing that really got her attention: It wasn't just technically possible. It was actually happening, at massive scale, right now. Just not in America.

Yet.

The 53 Million Americans Left Behind

Let's talk numbers for a second. **Fifty-three million US households cannot install rooftop solar.** Read that again. That's not a small niche. That's one-third of all American homes.

Why? Most live in apartments or condos where they don't own the roof. Others rent houses where the landlord isn't interested in solar. Some own homes with roofs that face the wrong direction, are too shaded, or are in such bad shape that adding solar panels would be like putting racing stripes on a car held together with duct tape.

For decades, these 53 million households watched their homeowning neighbors install solar panels, collect federal tax credits, slash their electric bills, and feel good about helping the environment. Meanwhile, renters and apartment dwellers got... nothing. Zero options. Complete exclusion from the solar revolution.

The message was clear: Solar is for homeowners. Period.

Except that message was wrong.

How Germany Accidentally Started a Revolution

In 2020, a German man named Klaus bought two solar panels at his local hardware store. Not a specialty solar shop. Not through an installer. The regular hardware store, right next to the garden supplies.

Klaus lived in an apartment. He mounted the panels on his balcony railing, plugged them into a standard wall outlet, and started generating electricity. No permits. No utility approval. No electrician. Just plug and play.

Germany's regulations didn't exactly encourage this. They just didn't explicitly prohibit it for small systems. And that tiny gap in the rules created something unexpected.

By the end of 2022, **Germany had 137,000 balcony solar systems installed**. That seemed like a lot. But then something happened. Energy prices spiked after Russia invaded Ukraine. Suddenly, electricity bills doubled or tripled. People got desperate for solutions.

In just the first six months of 2024, Germans installed **220,000 more balcony solar systems**. That's more installations in six months than in all previous years combined. By the end of 2024, **over one**

million German households were generating solar power from their balconies.

One million. In a country with 83 million people.

Do the math on that. Germany proved that roughly **8% of people want balcony solar badly enough to buy it within 12 months**. Another 9% said they're seriously considering it. That's not a niche market. That's mainstream adoption.

And here's what makes this even more remarkable: Germans aren't buying these systems primarily for environmental reasons. Yeah, they care about climate change. But the main driver is simple economics. These systems pay for themselves in 3-7 years, then keep saving money for another 20+ years.

The German government noticed this explosion of interest and did something smart. In 2024, they made it even easier. They raised the legal limit from 600 watts to 800 watts. They simplified registration requirements. They explicitly protected tenant rights to install balcony solar without needing landlord approval.

The result? The market accelerated even faster.

Utah Becomes America's Unlikely Solar Pioneer

Now let's jump back to the United States, specifically to Utah. Not exactly the state you'd expect to lead a green energy revolution, right? Conservative politics. Oil and gas industry. Not the obvious candidate.

But in March 2025, something surprising happened. **Utah's state legislature unanimously passed a bill creating a new category for "small portable solar generation devices" under 1,200 watts**. These systems would be exempt from interconnection requirements. No utility approval needed. No permits. No paperwork maze.

Unanimous. Not close. Not controversial. Every single legislator voted yes.

How did that happen in a red state? Simple. Republican State Representative Raymond Ward read a New York Times article about Germany's balcony solar boom. He realized this wasn't really an environmental issue. It was an economic freedom issue.

"This is about giving people the opportunity to save money and getting the government out of their way," Ward said. Hard to argue with that, regardless of your politics.

Utah became America's first state with clear, explicit legal protection for plug-and-play balcony solar. Not because of environmental activism. Not because of Democratic politics. Because it just made common sense.

And suddenly, other states started paying attention. Vermont began drafting similar legislation. Maryland and Pennsylvania expressed interest. The dominoes were starting to fall.

Why 2025 Is the Tipping Point

Here's why this moment matters. Actually, here are several reasons why this moment matters, and they're all converging right now:

The Regulatory Ice Is Breaking

For years, US regulations treated all solar installations the same. Want to plug in a 300-watt panel on your balcony? Same paperwork as a 10,000-watt rooftop system. Same interconnection agreements. Same utility approval process. Same permitting requirements.

That was insane. Like requiring a commercial driver's license to ride a bicycle.

But utilities had good reasons for caution. They worried about safety. They worried about grid stability. They worried about electricity flowing backward through old meters not designed for it. Some of those concerns were legitimate. Many were not.

The regulatory freeze is thawing. States are realizing you can write sensible rules that protect safety without making balcony solar impossible. Utah proved it. Other states are following.

The Technology Has Matured

Early balcony solar systems had issues. Inverters failed. Connections were sketchy. Monitoring was nonexistent. But ten years of European deployment solved most of those problems.

Modern systems include rapid shutdown features. They have anti-islanding protection that prevents backfeeding during grid outages. They come with monitoring apps that show exactly how much power you're generating. They use standardized connectors that work reliably.

The technology is ready. Proven. Safe.

The Economics Are Compelling

Here's the money part. In Europe, complete 400-watt balcony solar systems cost $325-550. In the US, prices are higher right now (around $400-800 for entry-level systems) because the market is still developing. But those prices will drop as volume increases.

These systems generate approximately 1-2 kilowatt-hours of electricity per day in decent sun conditions. That's enough to cover 10-30% of a typical apartment's electricity consumption. At average US electricity rates of $0.15-0.25 per kilowatt-hour, you're looking at annual savings of $100-300.

Do the math. A $500 system saving $200 per year pays for itself in 2.5 years. Then it keeps saving money for another 20+ years.

Oh, and there's a **30% federal solar tax credit** that applies to these systems (if they're permanently connected to your home's electrical system). That brings the effective cost down even further. But here's the urgency factor: Congress has proposed eliminating this credit by the end of 2025.

If you're thinking about balcony solar, you want to act before that credit disappears.

The Housing Crisis Meets the Energy Crisis

American renters are getting squeezed from both sides. Housing costs keep rising. More than half of all renters (51.8%) are "cost-burdened," meaning they spend over 30% of their income on housing. Many spend 40% or 50%.

At the same time, electricity costs are climbing. Utilities are raising rates. Climate change is driving up air conditioning demand. Electric bills that used to be $80 are now $150.

Renters are desperate for any way to reduce their monthly expenses. Balcony solar offers one of the few options they actually control. Your landlord can raise your rent, but they can't stop the sun from generating free electricity on your balcony (in most cases).

The Environmental Imperative

Yeah, we need to talk about climate change. Every ton of CO2 matters. But here's the thing about balcony solar: it's not asking people to sacrifice for the environment. It's asking them to save money and help the environment.

Each balcony solar system prevents about 559 kilograms of CO2 emissions annually. That's equivalent to 12% of what a car produces. Scale that across one million homes (like Germany did) and you've saved 559,000 metric tons of CO2. That's like taking 120,000 cars off the road.

Environmental solutions work best when they align with economic self-interest. Balcony solar does exactly that.

Real People, Real Results

Let me tell you about Marcus, who lives in a fourth-floor walkup in Queens. He installed a 400-watt balcony solar system in spring 2025. His south-facing balcony gets good sun from 9 AM to 4 PM most days.

Marcus spent $600 on his system after the federal tax credit. His electricity bill dropped from an average of $120 per month to $95 per month. That's $25 monthly savings, or $300 per year. His system will pay for itself in two years.

But the savings aren't the only benefit Marcus talks about. "It's the feeling of control," he says. "I'm generating my own power. My landlord can raise my rent, but he can't stop my solar panels from working. And when I move, I'm taking them with me."

That last part is huge. **Portability changes everything for renters.**

Traditional rooftop solar is permanently attached to a house you don't own. That's a non-starter. But balcony solar mounts on railings or uses freestanding brackets. When you move, you unplug it, unmount it, pack it up, and take it to your next apartment. Your investment moves with you.

Maria in Denver learned this firsthand. She installed balcony solar at her apartment in 2024. One year later, she moved to a different building. The entire process of dismounting, transporting, and reinstalling her system took three hours. Her investment was preserved. Her savings continued.

And here's something interesting: Maria's new apartment has a west-facing balcony instead of south-facing. Her power generation dropped slightly, but not as much as she feared. Even west-facing balconies can generate meaningful electricity with the right panel placement.

Breaking the Homeowner Monopoly

For too long, solar power has been a homeowner's privilege. A status symbol. Something that increases property values for people who already own property, while renters watch from the sidelines.

Balcony solar disrupts that monopoly.

It says: You don't need to own a house to generate solar power. You don't need a perfect south-facing roof. You don't need to take out a $20,000 loan. You don't need to live in one place for ten years to make the investment worthwhile.

You just need a balcony, $500-800, and a decent amount of sunlight.

That's not just a technological shift. It's a democratization of energy generation. It's economic power being distributed to people who've been excluded from the solar revolution.

And it's about damn time.

The Portable Power Advantage

Let's talk about what makes this different from every other home improvement. New countertops? Leave them behind when you move. Fresh paint? Stays with the apartment. New flooring? That's the landlord's now.

But your balcony solar system? That's yours. You bought it. You installed it. You benefit from it. And when your lease ends, you take it with you.

This portability solves the fundamental renter's dilemma: Why invest in a place you don't own?

Answer: Because this investment is portable.

Think about other portable electronics you own. Your laptop. Your TV. Your phone. You buy them, use them for years, and take them wherever you go. Balcony solar works the same way. It's a portable appliance that happens to generate electricity.

Some renters move every year or two. That's fine. Your balcony solar system will work just as well at your next apartment (assuming it gets decent sun). You're not locked in. You're not committed to one location. You maintain flexibility while enjoying immediate benefits.

That's a rare combination.

What's Different About This Moment

You might be thinking: If this technology has existed for years in Europe, why hasn't it come to America sooner?

Good question. The answer is bureaucracy, utility resistance, and regulatory inertia. US electrical codes were written decades ago for a centralized grid where power flows one direction: from utility to customer. The idea that customers might generate their own power and feed it back into the grid (even tiny amounts) freaked people out.

Safety concerns were real. Old mechanical meters could actually run backward if you generated more power than you consumed. That's a problem for utilities trying to bill accurately.

But here's what changed: Smart meters. Modern inverters. Proven safety standards from a decade of European deployment. The technical issues have been solved. The safety concerns have been addressed. The only thing left was political and regulatory will.

Utah provided that will in 2025. Other states are following. Federal standards are being developed. Manufacturers are entering the US market with proven products.

The dam is breaking.

This Book Is Your Roadmap

Look, I'm not going to pretend that balcony solar solves every problem. It doesn't. You can't power your entire apartment with 400 watts of solar panels. You can't completely eliminate your electric bill. You still need the grid for most of your electricity.

But 10-30% reduction in electricity costs? That's real money. For renters spending $120-200 per month on electricity, we're talking about $15-50 in monthly savings. That's $180-600 per year. Every year. For 20+ years.

And you get to do it on your terms. No landlord permission needed in most cases. No permanent modifications to the building. No multi-year commitment. Just clean energy and immediate savings.

This book exists because information about balcony solar in the United States is scattered, incomplete, and often wrong. Europeans have comprehensive guides. Americans have... fragments. Blog posts. Reddit threads. Scattered knowledge.

Not anymore.

In the chapters ahead, you'll learn exactly how balcony solar works. You'll understand the technology without needing an engineering degree. You'll navigate state regulations and figure out whether you need landlord permission. You'll discover how to maximize power generation even in shaded urban environments. You'll learn about battery storage options that let you use solar power at night. You'll calculate your specific ROI based on your city's electricity rates and sun exposure.

And you'll do all of this with real information, practical advice, and honest assessments of what works and what doesn't.

Because here's the truth: Balcony solar isn't a miracle solution. It's a practical tool that can meaningfully reduce your electricity costs while giving you a small measure of energy independence.

For 53 million American households, that's not nothing.

That's everything.

What You'll Take Away from This

So what's the bottom line here? What matters?

First, balcony solar is real, proven technology that works. Germany didn't install one million systems by accident. They did it because the technology delivers on its promises: meaningful electricity generation, reasonable payback periods, and genuine savings.

Second, America is finally catching up. Utah's 2025 legislation broke the regulatory logjam. Other states are following. The window is opening.

Third, the economics work for renters in ways traditional solar never could. Portable systems that you take when you move solve the fundamental investment problem of renting.

Fourth, you don't need permission to join the solar revolution anymore. You don't need a roof, a house, or a permanent address. You just need a balcony and enough determination to figure this out.

That's what this book is for. To turn that determination into action. To give you the knowledge, tools, and confidence to install your own balcony solar system and start generating your own electricity.

The revolution is here. The technology works. The regulations are changing.

Your move.

Chapter 2: Understanding Plug-and-Play Solar Technology

Here's what you need to know right upfront: Balcony solar isn't complicated. Actually, it's deliberately simple. That's the whole point.

If you can plug in a toaster, you can install balcony solar. If you can use a smartphone app, you can monitor your power generation. If you can follow basic assembly instructions, you can mount the panels.

This isn't rocket science. It's practical technology designed for normal people who don't have engineering degrees.

But (and there's always a but), you do need to understand what you're actually buying and how it works. Not because the technology is difficult, but because understanding it helps you make smart decisions. You'll know what features matter, which ones don't, and how to get the most value from your investment.

So let's break down exactly what balcony solar is, how it works, and what it can realistically do for your electricity bill.

What Is Balcony Solar, Really?

Balcony solar is a miniature solar power system designed to plug directly into your home's electrical system through a standard outlet. That's it. That's the fundamental concept.

Traditional rooftop solar requires electricians, permits, utility interconnection agreements, and permanent installation. Balcony solar requires... a plug. And a sunny spot.

A complete balcony solar system consists of four main components:

Solar panels convert sunlight into direct current (DC) electricity. These are smaller versions of the panels you see on rooftops. Most balcony systems use one to four panels, each producing 100-400 watts of power.

A microinverter converts that DC electricity into alternating current (AC) electricity, which is what your home actually uses. The inverter also regulates the power output to match your home's voltage and frequency.

Mounting hardware keeps your panels secured to your balcony railing, wall, or a freestanding frame. This varies depending on your specific balcony setup.

Cables and connectors link everything together and connect to your standard electrical outlet.

Some systems also include monitoring devices that track how much power you're generating, and battery storage that saves excess electricity for later use. We'll cover batteries extensively in Chapter 6. For now, focus on the basic system.

How Plug-In Solar Actually Works

Time for the step-by-step process. Don't worry, this is straightforward.

Step 1: Sunlight Hits the Panel

Your solar panel contains photovoltaic cells made of silicon. When sunlight hits these cells, it knocks electrons loose, creating a flow of electricity. This is called the photovoltaic effect, and it was discovered way back in 1839. The technology isn't new. It's just gotten much better and cheaper.

Step 2: DC Electricity Flows to the Inverter

The electricity generated by your panel is direct current (DC). That means electrons flow in one direction. Batteries use DC power. So does the 12-volt system in your car.

But your home uses alternating current (AC), where electron flow switches direction 60 times per second. AC is more efficient for transmitting power over long distances, which is why the electrical grid uses it.

So we need to convert DC to AC. That's the inverter's job.

Step 3: The Inverter Converts and Regulates

Modern microinverters are small, sophisticated devices. They don't just convert DC to AC. They also:

- Match the exact voltage and frequency of your home's electricity (120 volts, 60 Hz in the US)
- Monitor power quality
- Shut down automatically if they detect grid problems (this is called anti-islanding protection)
- Optimize power output based on sunlight conditions

Step 4: AC Electricity Enters Your Home

Here's where the magic happens. The AC electricity from your inverter flows through the cable and into your standard wall outlet. Once it enters your home's electrical system, it becomes part of your household circuit.

Now here's the clever part: Electricity always takes the path of least resistance. When your balcony solar system generates power, your appliances use that power first because it's closer and easier to access than electricity from the utility grid.

Think of your home's electrical system like a canal system with water flowing through it. The utility grid is pumping water in from one end. Your balcony solar system is also pumping water in, but from a different location. Your appliances (the outlets and devices) are drawing water out wherever they need it.

The water from your solar system gets used first because it's at a slightly higher pressure. Only when your solar system isn't generating enough power do your appliances start drawing from the grid.

Step 5: Your Meter Slows Down (or Stops)

When your appliances use solar-generated electricity instead of grid electricity, your utility meter slows down. In some cases, it stops completely during peak sun hours.

With newer smart meters, the utility company can track exactly how much electricity you're using versus generating. With older mechanical meters, they just spin slower (or not at all) when you're generating power.

Either way, you're using less electricity from the grid. That means lower bills.

Power Generation Reality Check

Okay, let's talk numbers. Real numbers, not marketing hype.

A typical 400-watt balcony solar system will generate approximately **1-2 kilowatt-hours (kWh) of electricity per day** under good conditions. That's the average. Some days more, some days less, depending on weather and season.

What does 1-2 kWh per day actually mean? Let's put it in context.

The average American apartment uses about **20-30 kWh per day**. Smaller studios might use 15 kWh. Larger apartments with multiple occupants might use 35-40 kWh.

So your 400-watt balcony solar system generating 1-2 kWh per day covers about **5-10% of your total electricity consumption**.

Hold on. I can hear you thinking: "That's not very much."

You're right. It's not very much. A 400-watt system won't power your entire apartment. It won't eliminate your electric bill. It won't make you energy-independent.

But here's what it will do: It will generate enough electricity to run your refrigerator continuously. Or power all your LED lights and charge your devices. Or run a small window AC unit for several hours. Or cover most of your baseline power consumption (the stuff that's always running: WiFi router, modem, cable box, refrigerator, etc.).

And here's the thing about that baseline consumption: It happens 24/7. Your refrigerator doesn't turn off at night. Your modem doesn't take weekends off. These devices consume electricity constantly, which means they account for a significant portion of your total bill.

If your balcony solar system covers that baseline consumption during daylight hours, you're saving money every single day the sun shines.

The 10-30% Offset Sweet Spot

Based on real-world data from European installations, most balcony solar systems offset **10-30% of household electricity consumption**. That's the realistic range.

Where you fall within that range depends on several factors:

System size matters. A 300-watt system will offset less than an 800-watt system. Obviously.

Your sun exposure matters. South-facing balconies in sunny climates generate more power than north-facing balconies in cloudy regions.

Your consumption patterns matter. If you're home during the day using electricity while your system generates power, you'll see higher offset percentages. If you're at work all day and consume most of your electricity in the evening, you'll see lower percentages (unless you add battery storage).

Your total consumption matters. If you live alone in a studio and use 15 kWh per day, a system generating 2 kWh per day covers 13% of your consumption. If you live with roommates in a larger apartment using 35 kWh per day, that same system only covers 6%.

The math is straightforward. More generation or less consumption means higher offset percentages.

Here's a real example: Tom lives in Phoenix. He has a 600-watt balcony solar system on his south-facing balcony. Phoenix gets excellent sun nearly year-round. Tom's system generates about 2.5-3 kWh per day.

Tom's apartment uses about 25 kWh per day (he keeps his AC running a lot in summer). His balcony solar offsets about 10-12% of his total consumption, or about $30 per month at Phoenix's electricity rates.

Tom's system cost $750 after the federal tax credit. At $30 monthly savings ($360 annually), his payback period is just over 2 years. After that, it's pure savings for another 20+ years.

That's the sweet spot. Meaningful savings without unrealistic expectations.

Microinverters vs. String Inverters

Let's talk about inverters for a minute, because this matters.

Microinverters are small inverters attached directly to each solar panel (or shared by two panels). Each panel operates independently.

String inverters are larger devices that handle multiple panels connected in series, like Christmas lights.

For balcony solar systems, **microinverters are almost always the better choice**. Here's why:

If one panel gets shaded or has issues, the other panels keep working at full capacity with microinverters. With string inverters, shading one panel reduces the entire system's output.

This matters a lot for balcony installations where partial shading is common. A neighboring building might shade part of your balcony in the morning. A tree might cast shadows in the afternoon. With microinverters, your unshaded panels keep generating maximum power.

Microinverters also make expansion easier. Want to add another panel later? Just plug it in. With string inverters, you need to reconfigure the entire system.

Most modern balcony solar systems come with microinverters. That's what you want.

Safety Features That Matter

Modern balcony solar systems include several important safety features. You need to understand these, not because you'll do anything with them, but because they explain why this technology is safe.

Rapid Shutdown: If you unplug your system, it stops generating power within milliseconds. No residual electricity. No danger. This feature is required by most electrical codes.

Anti-Islanding Protection: This is the safety feature that prevents your system from generating power during grid outages. Why does that matter? Because if the grid goes down and utility workers are trying to fix it, they need to be certain no electricity is flowing through the lines. Anti-islanding protection ensures your system shuts down automatically when it detects a grid outage.

Some people find this frustrating. "Why can't my solar panels power my home during a blackout?" they ask.

Answer: Safety. Without anti-islanding protection, you could accidentally electrocute a utility worker. So the system shuts down.

(If you want backup power during outages, you need battery storage with a transfer switch. That's a more complex setup we'll cover in Chapter 6.)

Ground Fault Protection: Detects electrical leaks and shuts down the system before anyone gets shocked.

Overcurrent Protection: Prevents too much electricity from flowing through your circuits.

These features work automatically. You don't do anything. But they're the reason balcony solar is safe, even for non-experts.

Why These Systems Can't Power Your Entire Apartment

Let's address the elephant on the balcony: Why so little power? Why only 300-800 watts? Why can't we just install bigger panels and power everything?

Several reasons.

Physical Space: Balconies are small. Most can fit 1-4 panels maximum. That limits total wattage.

Electrical Capacity: Standard outlets are rated for 15-20 amps in the US. That's about 1,800-2,400 watts maximum. Plug in an 800-watt solar system, and you're using only part of that circuit's capacity. Good. Safe. But it means you can't just keep adding panels infinitely.

Regulatory Limits: Most jurisdictions that allow plug-and-play solar limit systems to 600-1,200 watts. Utah allows up to 1,200 watts. Germany allows 800 watts. These limits exist partly for safety and partly to distinguish these simple systems from larger installations that require professional installation.

Grid Integration: Utilities are comfortable with tiny amounts of power backfeeding into the grid. They're less comfortable with larger amounts. The regulatory compromises that make plug-and-play possible include power limits.

Could you install more panels and more power? Sure. But then you'd need permits, utility interconnection agreements, professional installation, and all the complexity that balcony solar is designed to avoid.

The power limits are the price of simplicity.

Comparing Your Options

You've got three main ways to access solar power as a renter:

Balcony Solar (what this book is about): You buy and install a small system on your balcony. Costs $400-2,000. Generates 1-3 kWh per day. Portable when you move. No landlord permission needed in most cases.

Community Solar: You subscribe to a share of a large solar farm somewhere else. Your utility bill gets credited for your share of that farm's production. Costs vary (usually monthly subscription). You don't own anything. Not portable if you move out of the service area.

Convincing Your Landlord to Install Rooftop Solar: Good luck with that. Seriously, some landlords might be interested if you present a good business case. But you're asking them to invest $20,000-40,000 in a property you don't own. The benefits mostly go to them, not you. It's a tough sell.

Each option has merits. Community solar is great if it's available in your area and the subscription terms are favorable. Rooftop solar with landlord agreement is amazing if you can make it happen.

But balcony solar is the option you control. You don't need anyone's permission (in most cases). You don't need to hope someone else takes action. You just decide, buy, install, and start saving.

That control matters.

What You Need to Remember

Okay, let's wrap up the technical stuff with what actually matters.

Balcony solar is simple technology made from proven components. It's not experimental. It's not risky. It's straightforward physics and engineering that's been refined over decades.

The systems generate meaningful but limited power. Expect 1-3 kWh per day from a 400-800 watt system. That's enough to offset 10-30% of your consumption, which translates to real money saved.

Safety features are built in. Rapid shutdown, anti-islanding protection, and other safeguards mean you're not creating hazards by plugging solar panels into your apartment.

The technology is accessible. You don't need special skills or knowledge. If you can follow instructions and use basic tools, you can install balcony solar.

Realistic expectations lead to satisfaction. Don't expect to eliminate your electric bill. Do expect to reduce it meaningfully.

Don't expect energy independence. Do expect more control over your electricity costs.

The next chapter tackles the regulatory maze. Because understanding the technology is just the first step. Understanding the rules is what lets you actually implement this.

Chapter 3: State-by-State Regulatory Navigator

Let me tell you about Jennifer's experience. She lives in Sacramento. She bought a balcony solar system online. The kit arrived. She got excited. Then she started reading about regulations.

California requires interconnection agreements for systems that backfeed power to the grid. But wait, her system was designed not to backfeed. So did she need an agreement or not? She called her utility. They didn't know. Their rules were written for rooftop systems, not balcony solar. She called her city's building department. They didn't know either.

Jennifer spent three weeks trying to get a straight answer. Eventually, she found a nonprofit called Bright Saver that specialized in plug-and-play solar for California residents. They explained that if her system had specific monitoring equipment to prevent backfeeding, she didn't need an interconnection agreement. But she did need to notify her utility.

Three weeks of confusion for something that should have been simple.

That's the regulatory reality in America right now. Rules written for large rooftop systems being applied (or misapplied) to tiny balcony installations. Confusion everywhere. Nobody quite sure what's allowed and what isn't.

This chapter exists to cut through that confusion. You'll learn which states have clear rules, which are in gray areas, and how to navigate the mess in between.

The Utah Model: How One State Got It Right

Let's start with the success story.

In March 2025, **Utah passed HB 429, creating a legal category for "small portable solar generation devices" up to 1,200 watts**. The law specifically exempts these devices from interconnection requirements.

What does that mean in plain English? It means if you live in Utah and your balcony solar system is under 1,200 watts, you can plug it in without:

- Utility permission
- Interconnection agreements
- Special permits
- Grid connection fees
- Months of paperwork

You just install it and use it. The utility can't stop you. They can't charge you fees. They can't make you wait for approval.

That's revolutionary.

How did Utah pull this off? By making a simple distinction that other states have failed to recognize. There's a huge difference between:

A 10,000-watt rooftop system that might generate more power than the household uses, backfeeding significant electricity to the grid...

And a 600-watt balcony system that will never generate more power than the household's baseline consumption.

The first requires coordination with the utility. The second doesn't meaningfully affect the grid at all.

Utah's law includes one important requirement: Systems must comply with Underwriters Laboratories (UL) standards. We'll talk about UL certification in a minute. The short version is that reputable manufacturers already meet these standards.

What Utah's law doesn't require:

- Registration with the utility (though you may want to notify them anyway)
- Professional installation
- Special circuits or electrical work
- Inspections
- Fees of any kind

It's the plug-and-play model that works in Europe, finally codified into American law.

The result? Utah became the first US state where balcony solar is unambiguously legal and simple. Manufacturers are paying attention. Investors are paying attention. Other states are paying attention.

One state can change an entire industry.

States to Watch: The Momentum Is Building

Utah broke the ice. Now others are following.

Vermont introduced legislation in early 2025 similar to Utah's model. The bill would allow plug-and-play solar systems up to 800 watts without interconnection requirements. As of late 2025, the bill is moving through committee review.

Maryland is considering legislation that would streamline balcony solar installation in multi-family buildings. The focus is on tenant rights and preventing HOAs from blocking reasonable solar installations.

Pennsylvania lawmakers are drafting regulations specifically for micro-generation systems. The proposed framework would exempt systems under 1,000 watts from most interconnection requirements.

These aren't fantasies. These are real legislative efforts happening right now, inspired by Utah's success.

Why the sudden momentum? A few reasons.

Republican and Democratic legislators both like this issue. It's pro-consumer. It's pro-choice (in the "freedom to choose your energy" sense). It reduces electricity costs. It helps the environment. It's hard to find a principled reason to oppose it.

The German example provides proof of concept. Over one million installations with no significant safety incidents. That's hard to argue with.

Federal agencies are supportive. The Department of Energy has explicitly backed plug-and-play solar development. That gives state legislators confidence.

And there's growing recognition that current regulations are absurd. Requiring the same paperwork for a 400-watt balcony system and a 10,000-watt rooftop system makes no sense. States are realizing they can write better rules.

California's Unique Approach

California doesn't have plug-and-play legislation yet. But something interesting is happening anyway.

A San Francisco nonprofit called **Bright Saver** launched in January 2025 with a specific strategy: Design systems that comply with California's existing regulations by preventing backfeed to the grid.

Here's how it works. California law requires interconnection agreements for systems that export power to the grid. So Bright Saver's systems don't export power. They use monitoring equipment that continuously tracks household consumption and solar generation. If the system detects that it's about to generate more power than the house is consuming, it automatically throttles down power output.

Zero backfeed. Zero grid impact. Zero interconnection requirement needed.

Is this the ideal solution? No. The monitoring equipment adds cost and complexity. But it works within California's existing legal framework while advocates push for better regulations.

Bright Saver's initial systems are only 200-220 watts to ensure they never generate more power than typical baseline consumption (refrigerator, modem, etc.). That's smaller than ideal. But it's a start. And it's legal right now in California without permits or utility approval.

Other California companies are taking similar approaches. EcoFlow has released systems with smart monitoring that prevents backfeed. Jackery is developing California-compliant products. The market is adapting to regulatory constraints while working to change those constraints.

If you live in California, you have options. They're not as simple as Utah's model, but they exist.

Understanding Interconnection Agreements

Let's talk about the regulatory concept that causes most confusion: interconnection agreements.

An **interconnection agreement** is a contract between you and your utility company that governs how your power generation system connects to the grid. Traditional rooftop solar requires these agreements. They specify:

- Technical requirements for your system
- Safety standards you must meet
- Metering arrangements (how your generation is tracked)

- Compensation terms (how you get credited for excess power)
- Insurance requirements
- Inspection procedures

For rooftop solar, interconnection agreements make sense. These systems generate substantial power. They can backfeed thousands of watts to the grid. Utilities need to track this and ensure safety.

But for 400-watt balcony systems that barely dent household consumption? Interconnection agreements are overkill. Like requiring a commercial fishing license to use a goldfish bowl.

Most states haven't updated their laws to recognize this difference. So technically, you might need an interconnection agreement for a balcony solar system in states without specific plug-and-play legislation.

What does that mean practically?

Some utilities don't know or care. If your system is small and doesn't backfeed power, they may never notice or object. You're flying under the radar.

Some utilities have informal policies. They won't require agreements for systems under certain wattages, even though the official rules say they should. You need to call and ask.

Some utilities strictly enforce interconnection requirements. They'll insist on full agreements, even for tiny systems. This is frustrating but legal.

Your strategy depends on your specific utility. Before installing, call your utility company and ask:

"I'm considering installing a small plug-and-play solar system on my balcony, under 400 watts, that prevents backfeed to the grid. Do I need an interconnection agreement?"

Get their answer in writing if possible. Email is fine. You want documentation of what they told you.

If they require an interconnection agreement, ask about the process and timeline. Some utilities have streamlined processes for small systems that take just days. Others require months of back-and-forth. Know what you're getting into.

National Electrical Code: What Exists Now

The **National Electrical Code (NEC)** is a set of standards for safe electrical installation adopted by most US jurisdictions. It's updated every three years.

Here's the problem: The current NEC doesn't have specific standards for plug-and-play balcony solar. It was written assuming all solar installations would be permanently wired by licensed electricians.

Plug-and-play systems that literally plug into outlets aren't covered clearly. They exist in a gray area.

This creates confusion. Some jurisdictions interpret the NEC as allowing plug-and-play systems. Others interpret it as prohibiting them. Both sides are reading the same document and reaching opposite conclusions.

The good news? The next NEC revision cycle is developing specific guidance for plug-and-play systems. Industry groups, including North American Clean Energy, have formed working groups to propose appropriate standards.

What will these standards likely include?

- Power limits (probably 800-1,200 watts maximum)
- Anti-backfeed requirements or monitoring systems
- Plug and connector specifications
- Safety feature requirements (rapid shutdown, anti-islanding, etc.)

- Labeling requirements

Once these standards are adopted (likely in the 2026 NEC revision), jurisdictions will have clear guidance. The gray area will shrink dramatically.

Until then, you're navigating based on local interpretation.

UL Certification: Why It Matters

Underwriters Laboratories (UL) is an independent safety testing organization. Products with UL certification have been tested and meet specific safety standards.

For balcony solar, UL certification matters a lot. Here's the current situation:

Individual components (solar panels, inverters, cables) have UL standards and certifications. Reputable manufacturers already meet these.

Complete plug-and-play systems don't have a specific UL standard yet. UL is developing one (UL 7004 for plug-and-play solar systems), but it's not finalized as of late 2025.

This creates another gray area. Are complete systems without system-level UL certification safe? Generally yes, if the individual components are UL-certified. But some jurisdictions won't allow installation without complete system certification.

What should you do?

Buy from reputable manufacturers that use UL-certified components and follow industry best practices. Companies like EcoFlow, Anker, Jackery, and Bluetti are good bets.

Ask about certification status. Responsible manufacturers will explain which certifications they have and which are pending.

Avoid suspiciously cheap systems from unknown sellers. If a complete kit costs $200 when similar systems cost $600, there's probably a reason. That reason might be "corners were cut on safety."

Within a year or two, UL certification for complete systems will be standard. Manufacturers are already preparing. But in the interim, buyer caution is appropriate.

Local Utility Policies: The Wild West

Utilities vary wildly in their approach to balcony solar. Some are supportive. Others are hostile. Many are simply confused.

Friendly utilities recognize that tiny balcony systems don't meaningfully affect grid operations. They allow installation without complex procedures. They may even provide guidance and support.

Neutral utilities don't love balcony solar but won't fight it either. They'll apply their standard interconnection process, which might be reasonable or might be burdensome.

Hostile utilities see balcony solar as a threat to their business model. They'll enforce every possible requirement, delay approvals, charge high fees, and generally make your life difficult.

How do you know which type your utility is?

Research. Google "[your utility name] plug-and-play solar" or "[your utility name] micro-generation policy." See what comes up.

Check online forums. Reddit has active communities discussing balcony solar. Someone in your utility's service area has probably already asked these questions.

Call and ask directly. Utilities have customer service lines. Call and explain what you want to do. Gauge their reaction.

Look at their net metering policies. Utilities with generous net metering for rooftop solar tend to be more accommodating of balcony solar. Utilities fighting net metering tend to resist balcony solar too.

If you encounter a hostile utility, you have limited options. You can:

- Comply with their requirements (even if unreasonable)
- Work with advocacy groups to push for regulatory changes
- Install a system small enough to avoid their radar (risky)
- Wait for state legislation to override utility objections

It's frustrating. But it's the current reality in many areas.

HOA and Condo Association Regulations

Homeowners associations and condo boards add another layer of complexity. They might have rules prohibiting exterior modifications, regulating balcony appearance, or restricting solar installations.

Here's what you need to know:

Read your HOA documents thoroughly. Look for rules about balcony modifications, exterior appearance, and renewable energy systems. What exactly is prohibited? What's allowed?

Many states have solar access laws that limit HOA power to restrict solar installations. These laws were written for rooftop solar but may apply to balcony systems too. California, Colorado, Florida, and several other states have strong solar access protections.

Some HOAs are flexible. If rules prohibit permanent installations, emphasize that balcony solar is portable and temporary. If rules concern aesthetics, choose attractive, low-profile panels. Be willing to compromise on placement or appearance.

Document everything. Submit written requests. Get written responses. If the HOA approves, keep that approval letter. If they deny, you want documentation of their reasoning.

Consider collective action. If you get denied, other residents probably want solar too. Organize a group and petition the HOA board to change the rules. There's strength in numbers.

The good news? HOAs are generally less restrictive than they used to be. Solar panels are mainstream now, not exotic. Many boards recognize they can't block popular, environmentally-friendly technologies indefinitely.

But some boards are stubborn. Pick your battles carefully.

Building Code Variations by Metro Area

Building codes vary not just by state but by city and county. Some jurisdictions adopt the NEC verbatim. Others modify it extensively.

Major metro areas with relatively friendly codes:

Austin, Texas has streamlined permits for small renewable energy systems. Phoenix, Arizona has generally accommodating building departments. Portland, Oregon has progressive solar policies. Denver, Colorado has supportive regulations for renewable energy.

Major metro areas with complex codes:

New York City has extensive building regulations that can complicate balcony solar. San Francisco has multiple overlapping approval requirements. Boston has historic district restrictions that limit exterior modifications.

Your strategy depends on your location. Before purchasing equipment, contact your local building department and ask about requirements for balcony solar installation. Specifically ask:

- Do I need a permit for a plug-and-play solar system under 600 watts?
- What documentation do I need to provide?
- How long does the approval process take?
- What fees are involved?

Get this information upfront. Don't buy equipment and then discover you need a $500 permit and three months of approvals.

Some jurisdictions have no clue how to handle balcony solar. They've never dealt with it before. Be patient. Be willing to educate. Provide information about European standards and US states that have approved it. Sometimes you're blazing a trail.

Navigating the Gray Area Safely

Okay, let's be honest about the current situation. In many jurisdictions, the legal status of balcony solar is ambiguous. Not explicitly allowed. Not explicitly prohibited. Just unclear.

What should you do in a gray area?

Option 1: Wait for clarity. If your state is considering plug-and-play legislation, waiting might be smart. Once the law passes, installation will be straightforward.

Option 2: Seek formal approval. Contact all relevant authorities (utility, building department, HOA) and get explicit permission. This takes time and effort, but you'll have legal cover.

Option 3: Proceed cautiously without formal approval. Install a small system that doesn't backfeed power and doesn't violate any explicit rules. This is technically a gray area. Some people are comfortable with this. Others aren't.

I can't tell you which option is right for you. That depends on your risk tolerance, your specific location, and your circumstances.

What I can tell you is this: **Europe has proven these systems are safe**. Over one million installations with no significant safety incidents. The technology works. The risks are minimal. The barriers are regulatory, not technical.

Regulations will catch up. They always do. Technology moves faster than law. That's normal. If you're an early adopter, you navigate ambiguity. If you prefer clarity, you wait for the rules to settle.

Both approaches are valid.

What Matters Right Now

Here's the bottom line on regulations:

Utah has clear, simple rules. Other states are following. Within 2-3 years, most states will have plug-and-play provisions. We're early in that transition.

Until your state passes clear legislation, you navigate existing rules. That means understanding interconnection agreements, utility policies, building codes, and HOA restrictions.

The gray area is shrinking. Industry standards are being developed. Federal agencies are supportive. State legislators are acting. The regulatory environment is improving rapidly.

You're not breaking the law by installing balcony solar (in most cases). You might be in a gray area. You might need to navigate bureaucracy. But you're not a criminal for wanting to generate your own electricity.

The next chapter gets practical. We're done with regulations. Time to talk about actual installation, mounting options, and making balcony solar work in urban environments with shading, small spaces, and high-rise challenges.

Because understanding the rules is necessary. But implementing the solution is what actually matters.

Chapter 4: The Landlord Negotiation Playbook

Jason had his balcony solar system sitting in boxes in his living room for three weeks. Three weeks of staring at it. Three weeks of second-guessing.

The problem? His lease had a clause about "exterior modifications requiring written approval from management." Jason wasn't sure if balcony solar counted as an exterior modification. The panels would mount on his balcony railing with clamps. No drilling. No permanent changes. But still... exterior?

He drafted an email to his landlord. Deleted it. Drafted another. Deleted that too. He was stuck in permission paralysis, worried about asking and getting rejected, but also worried about not asking and getting in trouble later.

Finally, Jason called a tenant rights hotline. The advice he got changed everything: "Read your lease again. Look for what's actually prohibited, not what you think might be prohibited."

Jason read his lease. Really read it. The clause about exterior modifications specifically mentioned painting, structural changes, and permanent attachments. His solar panels were none of those things. They were portable equipment temporarily placed on his balcony. Like a barbecue grill. Or patio furniture.

Jason installed his system that weekend. No permission requested. No permission needed. His landlord noticed during a routine inspection two months later and said, "Oh, that's cool. Does it actually work?"

When You Actually Need Permission

Let's start with the legal reality. **You don't always need landlord permission for balcony solar.** Sometimes you do. Sometimes you don't. The difference matters.

You typically need permission if:

Your lease explicitly requires approval for solar installations. Some leases specifically mention solar panels, renewable energy systems, or electricity-generating devices. If yours does, you need permission. No way around it.

Your installation requires drilling, mounting, or permanent modifications. If you're drilling holes in walls, permanently attaching brackets, or making structural changes, that needs landlord approval in nearly all leases.

Your HOA or condo association has rules about balcony installations. These rules override your individual lease in many cases. Check both documents.

Local ordinances require landlord notification or approval. Some cities have rules requiring tenant notification to property owners for certain installations. Check your local codes.

You typically don't need permission if:

Your lease doesn't specifically prohibit portable solar equipment. Most leases don't mention solar panels at all because they were written before balcony solar existed. If it's not prohibited, it's generally allowed.

Your installation is genuinely portable and temporary. Clamp-on mounts. Freestanding frames. No drilling. No permanent changes. This is usually classified as portable personal property, like furniture.

You're not violating aesthetic or safety regulations. As long as your panels don't create safety hazards or egregiously violate building appearance standards, many landlords don't care.

Your state has solar access laws that protect your right. Some states limit landlords' ability to prohibit reasonable solar installations. These laws vary widely, but they exist.

Here's the key distinction many tenants miss: **There's a huge difference between "exterior modifications" and "portable equipment placed on exterior space."**

Modifications change the property. Equipment occupies space temporarily. Your landlord can restrict modifications. They usually can't restrict reasonable use of balcony space you're paying to rent.

Portable vs. Permanent Installations

This distinction is everything. Get this right and you might not need permission at all.

Portable installations:

- Use clamps, hooks, or weighted bases
- Require no tools for installation (or only basic tools)
- Can be removed in under an hour
- Leave no marks, holes, or damage
- Don't alter the balcony structure
- Can move with you to a new apartment

Permanent installations:

- Require drilling into walls or railings
- Use permanent brackets or mounts
- Need specialized tools for removal
- Leave holes or attachment points
- Modify the balcony structure
- Stay with the property when you leave

From a legal standpoint, portable installations are personal property. Permanent installations are property modifications. That distinction matters under most lease agreements and tenant law.

Your strategy is simple: **Choose the most portable mounting system that works for your balcony.** The more temporary and removable your installation, the less likely you need permission.

Some manufacturers specifically design "renter-friendly" systems with clamp mounts that require zero drilling. Companies like Robinsun and Sunflower Solar emphasize this feature. They know the rental market needs portable solutions.

The "No Structural Modifications" Argument

Most leases prohibit structural modifications. This makes sense. Landlords don't want tenants tearing down walls, changing plumbing, or altering the building.

But here's what many tenants and landlords don't understand: **Balcony solar panels clamped to a railing are not structural modifications.**

Let me explain why this argument works.

A structural modification changes the load-bearing capacity, integrity, or functionality of the building structure. Drilling holes in concrete walls? That's structural. Cutting through support beams? Definitely structural. Modifying the balcony railing system? Could be structural.

But clamping lightweight solar panels to an existing railing? That's not structural. You're not changing anything. You're using the existing structure as intended: to support items on the balcony.

Think about other things people put on balconies:

- Planters hanging from railings

- Privacy screens attached with clips
- Wind chimes hung from overhead structures
- String lights wrapped around railings

Nobody considers these structural modifications. They're temporary uses of balcony space. Solar panels with clamp mounts are exactly the same.

If your landlord claims balcony solar is a structural modification, you can politely but firmly explain the distinction. Bring photos showing the clamp mounts. Emphasize that nothing is being drilled, altered, or permanently changed. Show that removal takes minutes and leaves zero damage.

This argument works because it's legally accurate. You're not modifying structure. You're using rented space.

Building Your Business Case for Landlords

Okay, so maybe your lease does require permission. Or maybe you want to be proactive and get explicit approval. How do you convince your landlord to say yes?

Start by understanding landlord psychology. What do landlords care about?

Property value. They want to maintain or increase it.

Tenant satisfaction. Happy tenants renew leases and pay rent on time.

Avoiding problems. They don't want safety issues, code violations, or complaints from other tenants.

Minimizing maintenance. They don't want to deal with damage or repairs.

Legal compliance. They don't want liability or regulatory issues.

Your pitch needs to address these concerns while highlighting benefits. Here's the framework:

Property value argument: Solar panels are increasingly desirable amenities. Apartments with solar access appeal to environmentally-conscious renters. Some studies show green features increase property appeal and can justify slightly higher rents. Your installation makes the property more attractive to future tenants.

Tenant satisfaction argument: You're a responsible tenant who wants to stay long-term. You're investing in your living space, which shows commitment. Happy tenants are valuable. This makes you happy.

Risk mitigation argument: Your installation meets safety standards. The equipment is UL-certified (if it is). The installation is portable and causes no damage. You're providing insurance information (if applicable). You're eliminating all landlord liability.

Zero maintenance argument: You're responsible for all installation, maintenance, and removal. The landlord does nothing. Zero cost. Zero effort. Zero ongoing responsibility.

Legal compliance argument: Your system complies with local codes (verify this first). Some jurisdictions are developing solar access protections that could require landlords to permit reasonable installations. By approving now, the landlord avoids potential future conflicts.

Put this in writing. Don't make a casual verbal request. Draft a formal letter that reads professionally and addresses these points systematically.

Permission Request Templates That Work

Here's a template you can adapt:

Subject: Request for Approval of Portable Solar Panel Installation

Dear [Landlord/Property Manager Name],

I am writing to request approval for installing a small, portable solar panel system on my balcony at [Your Address]. I have researched this carefully and want to ensure full compliance with my lease agreement and your policies.

System Description:

- Two 400-watt solar panels with clamp-style mounting
- No drilling, screws, or permanent modifications required
- Complete installation and removal possible in under one hour
- Total weight: approximately 40 pounds
- All equipment meets UL safety standards

Benefits to Property:

- Increases property appeal to environmentally-conscious renters
- Demonstrates commitment to sustainability
- Creates no maintenance burden for property management
- Aligns with increasing demand for green living options

Safety and Liability:

- All equipment is certified for residential use
- Installation complies with [city/state] building codes
- My renter's insurance policy covers the equipment
- System includes safety features preventing grid backfeed
- No fire hazard or electrical safety concerns

Responsibilities: I will be fully responsible for:

- All installation and maintenance

- Any costs associated with the system
- Complete removal upon lease termination
- Ensuring no damage to property
- Compliance with all applicable regulations

Visual Impact: I've attached photos of similar installations that show the minimal visual impact. The panels will be positioned [describe placement] and maintain the aesthetic character of the building.

I would appreciate the opportunity to discuss this request with you. I'm happy to provide additional information, show you the equipment, or address any concerns you may have.

Thank you for considering this request. I look forward to your response.

Sincerely, [Your Name] [Contact Information]

Notice what this letter does:

- Uses professional, respectful language
- Provides specific details about the system
- Addresses potential landlord concerns proactively
- Emphasizes benefits and minimizes perceived risks
- Takes full responsibility for everything
- Offers to discuss and provide more information

This isn't a casual "Hey, can I put solar panels up?" request. It's a formal proposal that shows you've thought this through and you're being responsible.

What Your Lease Actually Says

Before doing anything, read your lease. Actually read it. Not skim it. Read every word of sections dealing with:

Modifications and alterations - What's prohibited? What requires approval? How is "modification" defined?

Use of premises - Are there restrictions on how you can use your balcony? Are solar panels specifically mentioned?

Exterior installations - What rules govern things attached to or placed on exterior spaces?

Utilities and electrical systems - Are there restrictions on electricity generation or consumption?

Insurance and liability - What are your responsibilities for property damage or accidents?

Removal upon termination - What must be removed when you leave? What condition must the property be in?

Look for the specific language. Words matter. A lease that prohibits "permanent structural modifications" is different from one that prohibits "all exterior attachments." The first probably allows portable solar. The second might not.

Many leases include standard language about "reasonable use of premises." This is helpful. Generating solar power for your own use is arguably reasonable. It's quiet, safe, and doesn't interfere with other tenants.

Some leases include clauses about "nuisance or annoyance to neighbors." Solar panels don't create noise, odors, or disruption. Not a nuisance.

If your lease is silent on solar panels (most are), you're in a gray area. Gray areas favor tenants in most jurisdictions, assuming you're not causing damage or violating explicit rules.

Tenant Rights You Should Know

Several legal principles protect tenants who want to install solar:

Quiet enjoyment - You have the right to use your rented space without unreasonable interference. Installing portable equipment on your balcony is generally part of quiet enjoyment.

Solar access laws - Some states have laws limiting landlords' ability to prohibit solar installations. California, Colorado, and Florida have strong protections. Many other states have weaker but still relevant provisions.

Discrimination protections - Landlords can't arbitrarily allow some tenants to install solar while denying others without legitimate reasons.

Reasonable accommodation - If you can demonstrate that solar reduces your electricity costs significantly and you're cost-burdened, some jurisdictions require landlords to make reasonable accommodations.

State-specific provisions - Some states are developing explicit tenant rights around distributed energy resources. Check your state's recent legislation.

Now, here's the reality: Most landlords don't know these laws. Most tenants don't either. Information asymmetry favors whoever understands the rules better.

If your landlord denies permission unreasonably, consult a tenant rights organization or attorney. Sometimes a lawyer's letter explaining relevant laws changes a landlord's mind quickly.

Germany's Tenant Privilege Approach

Germany took a different approach that's worth understanding, even though we can't directly apply it in the US yet.

In 2024, German law established that **balcony solar is a privileged tenant right**. Tenants don't need landlord permission for systems meeting certain criteria:

- Under 800 watts
- No permanent modifications
- Meets safety standards
- Properly registered (simple online process)

Landlords can object only if there are specific, documented safety concerns or aesthetic issues that substantially harm the property. "I don't like it" isn't sufficient reason.

This approach recognizes that balcony solar is in the public interest (climate, energy independence, cost savings) and shouldn't be subject to arbitrary landlord veto.

The US doesn't have this yet. But several state legislators have cited the German model as they draft balcony solar legislation. Utah's law moves in this direction. Others are following.

The point? This is the future. The question isn't whether tenant solar rights will expand. The question is how quickly.

Insurance and Liability Considerations

Some landlords worry about liability. "What if your solar panel falls and hits someone? What if it causes a fire? Am I responsible?"

Address these concerns directly.

Your renter's insurance should cover your personal property, including solar panels. Call your insurance company and ask. Most policies cover portable solar equipment without additional premiums. Get written confirmation.

Landlord insurance typically doesn't need to change. Your portable solar system is your property, not theirs. They're not liable for your belongings.

Product liability insurance is the manufacturer's responsibility. If the equipment is defective and causes damage, the manufacturer is liable, not you or your landlord.

Installation liability is yours. If you install carelessly and cause damage, you're responsible. But proper installation eliminates this risk.

Offer to add your landlord as an "additional insured" on your renter's policy for solar-related incidents. This costs nothing and gives landlords peace of mind.

Some tenants get written documentation from their insurance company confirming coverage and send that to their landlord with the permission request. This proactive approach eliminates the "what if something happens" objection.

Documentation That Protects You

Whether you get explicit permission or proceed without it, documentation matters.

Photograph everything before installation. Show the current condition of your balcony. Any existing scratches, marks, or damage. Date-stamped photos are evidence that you didn't cause pre-existing issues.

Keep all equipment documentation. Receipts, manuals, safety certifications, warranty information. This proves you used quality equipment properly.

Document your lease analysis. Write down which lease clauses you reviewed and why you concluded permission wasn't required (if applicable). This shows good faith effort to comply.

Save all communications with landlord. Every email, every text, every letter. If disputes arise later, you have records.

Document installation process. Photos of proper installation. Evidence that you followed manufacturer instructions. Proof of care and professionalism.

Get move-out inspection documentation. When you leave, make sure the balcony condition is documented. No damage means no security deposit deductions.

This seems like overkill. But documentation protects you if anything goes wrong. Most of the time, nothing goes wrong. But when it does, having records is the difference between losing your security deposit and getting it back.

When Your Landlord Says No

So you asked. They said no. Now what?

First, understand why. Ask for specific reasons. "I don't want solar panels" isn't a reason. "I'm concerned about damage" is a reason you can address. "It violates building codes" is a factual claim you can verify or dispute.

Second, address specific concerns. If they worry about damage, offer to pay a small additional deposit. If they worry about aesthetics, show photos of attractive installations. If they worry about liability, provide insurance documentation.

Third, offer compromises. Maybe they'll approve a smaller system. Maybe they want specific mounting hardware. Maybe they'll approve for a trial period. Find middle ground.

Fourth, escalate if appropriate. If the landlord is a property management company, appeal to the owner. If you're in an HOA, appeal to the board. Sometimes higher-ups are more reasonable.

Fifth, consult legal resources. Tenant rights organizations can advise whether the denial is legally valid. Some denials are arbitrary and unenforceable. Legal advice clarifies your options.

Sixth, accept the situation or move. Sometimes landlords are unreasonable and legally allowed to be. If solar is really important to you, find a more accommodating landlord when your lease ends.

But here's the thing: Most landlords don't say no if you present the request professionally and address their concerns proactively. The majority just don't care as long as you're not causing problems.

Organizing with Other Tenants

Want to increase your chances of approval? Don't request alone. Organize.

If five tenants request permission for balcony solar simultaneously, the landlord pays more attention. Multiple requests demonstrate genuine demand. It shows this isn't one weird tenant with an oddball hobby. It's a movement.

Collective requests also share work. One person drafts the letter. Everyone signs. One person researches insurance. Everyone benefits. One person finds product recommendations. Everyone orders together.

Building-wide solar initiatives have additional benefits. You can negotiate group purchasing for better prices. You can share installation expertise. You can create a community of solar users who help each other troubleshoot issues.

Some apartment complexes have formalized this. A solar committee proposes building-wide policies that allow balcony solar under certain conditions. Management approves the policy once, and all tenants can participate without individual approval processes.

This approach works best in tenant-friendly buildings with engaged residents. But even in less ideal situations, numbers matter. Three requests are easier to deny than twenty.

What Actually Happens Most of the Time

Let me tell you the reality based on what's happening in Europe and early US adopters:

Most landlords don't notice or care. Seriously. You install portable solar panels on your balcony. They never come up. Your lease renews. You keep using them. No issues.

When landlords notice, most are curious, not hostile. "Oh, do those work? How much do they save you?" is the typical reaction. Anger or immediate prohibition is rare.

Denials usually come from property management companies, not individual landlords. Corporate policies are more rigid than individual judgment. Individual landlords are more flexible.

Good tenants get more leeway. If you pay rent on time, don't cause problems, and take care of the property, landlords give you more freedom. Good relationships earn flexibility.

Precedent matters. Once one tenant installs balcony solar successfully, others can point to that precedent. "You allowed Unit 3B to install solar. I'm requesting the same."

The fear of landlord rejection is often worse than the reality. Most conflicts never happen. Most requests get approved. Most landlords ultimately don't care about something that doesn't affect them.

But you can't rely on indifference. You need to know your rights, understand your lease, and be prepared to advocate for yourself.

Action Plan

Here's your action plan:

1. Read your lease thoroughly
2. Determine if permission is legally required
3. Document everything
4. Choose portable equipment
5. Draft a professional request if needed
6. Address landlord concerns proactively
7. Proceed with installation appropriately

Sometimes this takes weeks. Sometimes it takes days. But the process isn't mysterious. It's straightforward legal analysis combined with good communication.

You're not asking for something unreasonable. You're requesting permission to generate clean energy and reduce your electricity costs using equipment that causes no damage and creates no problems.

That's not radical. That's sensible.

And increasingly, landlords, laws, and society agree.

What You Need to Take Away

Permission isn't always required. Understand the legal distinction between modifications and portable equipment.

Documentation protects you. Photograph everything. Keep records. Get things in writing.

Professional requests get better results. Address concerns proactively. Offer solutions. Take responsibility.

Your rights are stronger than you think. Tenant protections exist. Solar access laws are expanding. Legal trends favor you.

Most landlords are reasonable. The horror stories are outliers. Most situations resolve positively.

The next chapter tackles actual installation. Because understanding landlord negotiations is necessary. But installing your system is what actually generates electricity and saves money.

Chapter 5: Urban Installation Mastery

David's balcony faces north. Straight north. The worst possible direction for solar panels in the Northern Hemisphere.

He almost gave up on balcony solar entirely. All the guides said "south-facing balconies are ideal" and "north-facing balconies won't work." Dead end, right?

But then David found a forum post from someone in Berlin who had the same problem. North-facing balcony. Still installed solar panels. Still seeing results, just reduced output.

The Berlin resident explained: "North-facing panels get indirect light, morning and evening sun from the sides, and reflection from nearby buildings. My system generates about 40% of what a south-facing system would. But 40% of something is better than 100% of nothing."

David did the math. His electricity bill was $140 monthly. A south-facing system might save him $35 monthly. His north-facing system might save him $14 monthly. Over twenty years, that's still $3,360 in savings from a $600 investment.

He installed anyway. His actual results? About $18 monthly savings. Better than the conservative estimate. His panels caught more reflected light and indirect sun than expected.

The lesson? **Optimal isn't the only option. Suboptimal can still be worthwhile.**

Why Direction Matters

Let's talk about solar panel orientation and why it affects power generation so much.

The sun travels from east to west across the southern sky (in the Northern Hemisphere). Solar panels generate maximum power when pointed directly at the sun. Simple physics.

South-facing panels catch direct sunlight from mid-morning through mid-afternoon. They get the longest period of direct sun and the highest sun angles at midday. This produces maximum power. South is the gold standard.

East-facing panels get strong morning sun from sunrise until about noon. As the sun moves west, power generation drops. East-facing systems generate 70-85% of what south-facing systems produce, with most power generated in the morning.

West-facing panels are the mirror image. Low power in the morning, peak generation in the afternoon and early evening. Also 70-85% of south-facing output, but the power comes at different times.

North-facing panels get no direct sun (in the Northern Hemisphere). They rely on indirect light, reflection, and scattered sunlight. Output drops to 30-50% of south-facing systems.

But here's what the simple guides miss: **Your consumption patterns matter as much as your generation patterns.**

If you work from home and use electricity all day, south-facing panels align well with your consumption. If you're gone all day and use most electricity in the evening, west-facing panels might actually be better for you despite lower total output. The power generation matches when you actually need it.

David's north-facing panels generate less power total. But they generate steady, moderate power throughout the day, which perfectly matches his constant baseline consumption (refrigerator, modem, etc.).

Making Peace with Suboptimal Orientation

Most apartment balconies face whatever direction the building faces. You don't get to choose. So what do you do when your balcony isn't south-facing?

Option 1: Install anyway and accept reduced output. Run the numbers. Calculate your expected savings at 40-85% of optimal output. If the payback period is still acceptable (under 7-8 years), install. Reduced savings are still savings.

Option 2: Angle your panels creatively. Some mounting systems allow panels to face different directions than your balcony. For example, corner-mounted panels on an east-facing balcony can angle southeast to catch more midday sun. This requires more sophisticated mounting but can increase output significantly.

Option 3: Use bifacial panels. These panels generate electricity from both sides. The rear side catches reflected light from your balcony floor or walls. On less-than-ideal balconies, bifacial panels can generate 10-20% more power than standard panels.

Option 4: Maximize panel area. If you can't optimize direction, maximize quantity. Two panels with suboptimal orientation might generate similar total power to one optimally-oriented panel.

Option 5: Wait for better technology. Panel efficiency improves every year. Future panels will generate more power from indirect light. If your balcony is severely suboptimal and you're not in a hurry, waiting might make sense.

Most people choose Option 1. Install anyway. Accept suboptimal results. The investment still pays off, just slower.

Shading Analysis Tools You Can Use

Urban environments mean shading. Lots of shading. Buildings, trees, other structures blocking sun at different times of day.

Professional solar installers use sophisticated tools for shading analysis. Solmetric SunEye devices cost thousands of dollars. That's not reasonable for balcony solar.

But you can do basic shading analysis with free tools:

Sun Seeker App (smartphone): Uses your phone's camera and GPS to show the sun's path across the sky for any date. Stand on your balcony and see exactly when sun hits and when shade appears. Cost: free for basic version, $10 for premium.

Google Earth Pro (desktop): Shows 3D building layouts. You can simulate sun angles for different dates and times to see what casts shadows on your balcony. Free software.

The Shadow App (smartphone): Specifically designed for shadow analysis. Shows how shadows change throughout the day and year. About $2.

The DIY Method: Spend an entire sunny day observing your balcony. Take photos every hour showing sun and shade patterns. Do this on both a summer day and winter day if possible. Free, but time-intensive.

What are you looking for?

Unshaded hours per day: How many hours of direct sunlight does your balcony get? More is better, but even 4-5 hours is workable.

Shade patterns: Does shade come from one direction or multiple? Consistent shade from one direction is easier to work with than shifting shade from multiple sources.

Seasonal variation: Does your balcony get more sun in summer than winter? This is normal, but extreme variation affects annual production.

Reflected light: Do nearby buildings or surfaces reflect additional light onto your balcony? This can boost power generation on otherwise shaded balconies.

Professional installers aim for 90%+ sun availability. For balcony solar, 60-70% sun availability can still work. Run the financial numbers and decide if it makes sense.

Working with Partial Shade

Partial shade is the enemy of solar power. But it's also the reality of urban balconies.

Here's why partial shade causes problems: When one part of a solar panel is shaded, the entire panel's output drops dramatically. A panel that's 20% shaded might lose 80% of its power output. That seems wrong, but it's how solar panels work.

The solution? **Microinverters and careful panel placement.**

Microinverters (we talked about these in Chapter 2) let each panel operate independently. If one panel is shaded, the others keep producing full power. This is critical for urban installations with partial shade.

Panel placement strategy for partial shade:

1. Identify the least-shaded area of your balcony
2. Place panels there first
3. If adding multiple panels, separate them to minimize simultaneous shading
4. Avoid placing panels where shade is consistent (permanent shade from adjacent buildings)
5. Focus on areas that get morning sun, midday sun, or afternoon sun—pick the best times

Consider adjustable mounts. Some mounting systems let you tilt or rotate panels seasonally. Summer sun is higher; winter sun is lower. Adjusting panel angle with seasons can minimize shade impact.

Use bypass diodes. Quality solar panels include bypass diodes that route power around shaded cells. This doesn't eliminate shade losses but reduces them. Make sure your panels have bypass diodes (most modern panels do).

Accept imperfection. Partial shade reduces output. That's reality. But 60% output on a partially-shaded balcony is still generating electricity and saving money.

Balcony Types and Mounting Solutions

Not all balconies are created equal. Your mounting strategy depends on what you're working with.

Railing-Mount Balconies

Most balconies have railings. These are ideal for clamp-style mounts that require no drilling.

Advantages:

- No permanent modifications
- Easy installation and removal
- Adjustable positioning along the railing
- Good angle options for optimal sun exposure

Challenges:

- Weight limits on railings
- Railing design must accommodate clamps
- Wind load concerns on upper floors
- Aesthetic visibility from street

Best mounting hardware: Adjustable clamps that fit various railing diameters and styles. Look for mounts with rubber padding to prevent scratching and improve grip.

Wall-Mount Balconies

Some balconies have solid walls instead of railings. These require brackets attached to walls.

Advantages:

- Stable mounting surface
- Can handle heavier panel arrays
- Less visible from street level
- Better wind resistance

Challenges:

- May require drilling (check lease)
- Less flexibility in positioning
- Harder to adjust angles
- Need to find studs or use appropriate wall anchors

Best mounting hardware: L-brackets with tilt adjustment. If drilling is required, use proper concrete anchors or mount into studs.

Freestanding Balconies

Balconies where you can't (or don't want to) attach anything to existing structures.

Advantages:

- Zero building attachment
- Ultimate portability
- Easy repositioning
- No landlord concerns about modification

Challenges:

- Weighted bases needed (wind resistance)
- Takes up floor space
- Less stable in high winds
- May be tripping hazard

Best mounting hardware: Freestanding frames with broad, weighted bases. Some systems use fillable bases (add water or sand for weight, empty for moving).

Juliet Balconies (small balconies, sometimes just window railings)

These tiny spaces are common in older buildings and studio apartments.

Advantages:

- Often south-facing (buildings designed for light)
- Still better than no balcony at all

Challenges:

- Very limited space (maybe 1 panel maximum)
- Safety concerns with installation
- May not support much weight
- Building aesthetic concerns

Best approach: Single small panel (100-200 watts) with minimalist mounting. Focus on systems designed specifically for minimal space.

Maximizing Limited Square Footage

Small balconies require creative thinking. Standard 400-watt panels measure about 6 feet by 3 feet. If your balcony is only 8 feet by 4 feet, one panel dominates the space.

Strategies for small balconies:

Vertical mounting: Instead of tilting panels toward the sun (which takes more space), mount them vertically on walls or railings. Output drops 10-20%, but space efficiency increases dramatically.

Flexible panels: Newer thin-film and flexible panels can mount on curved surfaces or uneven walls. Some can even be temporarily adhered to balcony walls or floors.

Folding systems: Some balcony solar systems use folding panels that can be opened when needed and folded flat when you want full balcony use. More expensive but solves the space problem.

Overhead mounting: In some cases, panels can mount overhead using ceiling brackets or cable systems. This requires structural assessment but frees floor space.

High-efficiency panels: Pay more for high-efficiency panels (22-24% efficiency) that generate more power per square foot. A 400-watt high-efficiency panel is smaller than a 400-watt standard panel.

Accept smaller systems: Maybe you can only fit 200-300 watts. That's okay. Smaller savings are still savings. Don't let perfect be the enemy of good.

High-Rise Challenges Nobody Talks About

Installing balcony solar on the 15th floor is different from installing on the 2nd floor. Height creates unique challenges.

Wind load increases with height. Wind speeds are significantly higher on upper floors. A mounting system that's stable on the third floor might get ripped off on the twentieth floor.

Solution: Use mounts with higher wind ratings. Add extra securing points. Consider freestanding weighted systems instead of railing clamps. Some manufacturers rate their mounts for specific maximum floors.

Access for installation is harder. Carrying solar panels up to the 15th floor (if your elevator is broken) is no joke. Panels are awkward and fragile.

Solution: Schedule installation when you can use service elevators. Get help. Use protective padding when transporting panels through tight spaces.

Building sway affects upper floors. Tall buildings move in wind. This movement can stress mounting hardware over time.

Solution: Use flexible mounting systems with some give. Check and tighten all connections every few months.

Fire department access. Some cities require balconies on upper floors to remain clear for emergency access. Check local fire codes.

Solution: Use removable systems that can be quickly demounted if needed. Document that your installation meets clearance requirements.

Falling object concerns. If a panel or component comes loose, it falls farther and causes more damage.

Solution: Overengineer your mounting. Use safety cables or secondary attachment points. Insurance is definitely recommended.

Fire Escape Regulations

Some balconies serve as fire escape routes. Others connect to fire escapes. Solar installations must not block emergency egress.

General rules (verify locally):

36-inch clearance is typically required for emergency exit paths. Your panels and mounting equipment can't narrow walkways below this width.

No blocking doors or windows that serve as emergency exits. Panels can't obstruct opening doors or prevent climbing through windows.

Quick removal capability may be required. Fire departments want to know that your installation can be removed quickly in an emergency (though this is rare in practice).

Fire-resistant materials may be required for mounting hardware. Check that your mounting system uses fire-rated materials if codes require it.

Before installation on balconies with fire escape functions, contact your local fire department's inspection division. Ask about requirements. Get guidance in writing if possible. Fire codes are serious; violations can result in forced removal and fines.

Most standard balconies that aren't part of fire escape routes have no special fire-related restrictions on solar panels. But verify for your specific situation.

Mounting Systems Compared

Let's break down the main mounting options with pros, cons, and costs.

Clamp Mounts ($30-80 per panel)

Clamp onto balcony railings without drilling. Most popular option for renters.

Pros: No drilling required, easily removable, adjustable positioning, works on most railing types

Cons: Limited to balconies with railings, wind load concerns, may scratch railings without padding

Best for: Renters with standard railing balconies, temporary installations, anyone avoiding permanent modifications

Bracket Mounts ($50-150 per panel)

Bolt to walls or solid surfaces. Requires drilling.

Pros: Very stable, handles higher wind loads, precise angle adjustment, professional appearance

Cons: Requires drilling holes, less portable, may void security deposits, needs landlord permission

Best for: Owners or tenants with explicit permission, long-term installations, high-wind areas

Weighted Freestanding ($100-250 per system)

Ground-based frames with weighted bases. No attachment to building.

Pros: Zero building attachment, completely portable, flexible positioning, easily removed

Cons: Takes up floor space, expensive, heavy to move, may tip in extreme wind

Best for: Tenants forbidden from attaching to building, temporary setups, renters wanting maximum flexibility

Adhesive Mounts ($20-60 per panel)

Strong adhesive pads bond panels to smooth surfaces temporarily.

Pros: No drilling, works on smooth walls, low profile, inexpensive

Cons: Surface must be clean and smooth, may damage paint upon removal, not suitable for textured surfaces, temperature-sensitive adhesion

Best for: Smooth-walled balconies, lightweight panels, short-term installations

DIY Custom Mounts ($10-50 materials)

Build your own mounting system from hardware store materials.

Pros: Customizable to unusual balcony configurations, inexpensive, satisfying if you enjoy DIY

Cons: Requires skills and tools, no engineering certification, potential safety issues, trial and error

Best for: Experienced DIYers, unusual balcony configurations, budget-conscious installers with skills

For most renters, clamp mounts are the obvious choice. They balance portability, cost, and ease of installation.

DIY Installation Step-by-Step

Ready to actually install? Here's the process for a typical two-panel system with clamp mounts. Total time: 2-4 hours.

Materials needed:

- Solar panels (2x)
- Microinverter(s)
- Mounting clamps (2-4 sets)
- MC4 connectors and cables
- Plug-in cable for outlet connection
- Adjustable wrench or socket set
- Rubber pads (if not included with clamps)
- Safety tether or cable (optional but recommended)

Step 1: Pre-Installation Prep

Check the weather. Install on a calm, dry day. Wind makes panel handling difficult. Rain is obvious bad idea.

Read all manufacturer instructions completely before starting. Watch installation videos if available.

Clear your balcony. Give yourself workspace. You'll be moving panels around, attaching hardware, routing cables.

Test your outlet with a multimeter or outlet tester. Confirm it's working properly. You need a functioning outlet for your system.

Step 2: Dry-Fit Everything

Before securing anything permanently, lay out all components on your balcony. Position panels where you think they'll go. Check that cables reach. Verify that you have all parts.

This dry-fit catches problems before you've tightened anything. Much easier to adjust now than after installation.

Check sun exposure at your chosen positions. Remember the shading analysis from earlier? Now apply it.

Step 3: Mount the First Panel

Attach mounting clamps to your panel frame. Most clamps slide onto the panel edge and tighten with screws or bolts.

Position the panel on your railing. Level it (you can eyeball this; perfection isn't required).

Tighten the clamps gradually. Don't overtighten—you'll crack the panel or damage your railing. Snug is enough.

Add rubber padding between clamps and railing if not already present. This prevents scratching and improves grip.

Rock the panel gently. It should be stable but not immovable (you'll want to adjust if needed).

Step 4: Mount the Second Panel (if applicable)

Repeat the process for your second panel. Try to keep panels at similar angles and orientations.

If using a single microinverter for two panels, position panels close enough that cables reach comfortably. Don't strain cables.

Step 5: Connect the Microinverter

Attach the microinverter to your mounting system or a secure location on the balcony. Some inverters mount on panel frames. Others need separate mounting.

Connect solar panels to inverter input using MC4 connectors. These connectors click and lock. Push until you hear/feel the click.

Make sure connections are secure. Loose connections create resistance, heat, and potential failure.

Step 6: Run Cables Safely

Route cables along balcony railings or walls. Avoid creating trip hazards across walking areas.

Use cable ties or clips to secure cables. Cables should not dangle loosely or be under tension.

Protect cables from weather exposure as much as possible. Some systems include cable conduits or covers.

Step 7: Connect to Outlet

Your system includes a plug-in cable from the microinverter to your outlet. Plug it in.

The inverter should power on (usually indicated by lights or sounds). Check the manufacturer's manual for what to expect.

Step 8: Verify Operation

Use your monitoring system (if included) to confirm the panels are generating power. On a sunny day, you should see immediate generation.

If no monitoring system, use a power meter on the outlet to confirm your system is feeding electricity into your circuit.

Check that your utility meter is slowing down or stopping during peak sun hours. This confirms the system is working.

Step 9: Safety Check

Rock each panel gently again. Everything should be secure.

Check all electrical connections. Nothing should be loose or exposed.

Verify that cables aren't pinched, strained, or damaged.

Make sure the installation doesn't create trip hazards, block pathways, or violate safety clearances.

Step 10: Document Everything

Take photos of your installation from multiple angles. Document that it's properly installed and causes no damage.

Keep all instructions, warranties, and receipts organized.

Note your installation date and any relevant information (serial numbers, etc.).

Done. You're now generating solar power.

When to Hire an Electrician

Most plug-and-play balcony solar systems don't require electricians. That's the whole point—DIY accessibility.

But sometimes professional help makes sense:

Your outlet is questionable. If your outlet is loose, damaged, or you're not sure it's properly grounded, hire an electrician to inspect/upgrade it before connecting solar equipment.

You want a dedicated circuit. Instead of plugging into an existing outlet, you want a new circuit specifically for your solar system. This requires professional electrical work and permits.

Your building requires professional installation. Some buildings or HOAs mandate professional installation for insurance or liability reasons. Check requirements before starting.

You're uncomfortable with electrical work. If the idea of plugging things in makes you nervous, hire someone. Safety first.

You want system optimization. Electricians familiar with solar can optimize your setup, ensure code compliance, and handle complications you might not anticipate.

Your system is more complex. If you're adding battery storage, smart monitoring, or other advanced components, professional installation is often worth the cost.

Cost for professional installation varies widely ($200-800 typically). Get quotes from electricians experienced with solar installations.

But for standard plug-and-play systems under 600 watts, most people successfully install themselves. The manufacturers design these systems for DIY installation. Trust the process.

Seasonal Optimization Strategies

The sun's position changes dramatically between summer and winter. Some mounting systems let you adjust panel angles seasonally to maximize output.

Summer optimization: Higher sun angle means panels should be tilted less steeply (or mounted flatter). In summer, aim for tilt angles equal to your latitude minus 15 degrees.

Winter optimization: Lower sun angle means panels should be tilted more steeply. In winter, aim for tilt angles equal to your latitude plus 15 degrees.

If you adjust seasonally: Change angles twice per year, typically in late October/early November (winter angle) and late March/early April (summer angle).

If you don't adjust: Set one angle and leave it year-round. The "optimal" static angle is approximately equal to your latitude. This balances summer and winter performance.

Many balcony solar systems use fixed mounting and don't offer angle adjustment. That's fine. The annual production difference between seasonal adjustment and fixed mounting is typically 5-15%. Meaningful but not critical.

If your mounting system allows adjustment, do it. If not, don't worry about it. Fixed mounting still works well.

Keeping Neighbors and Management Happy

Your installation exists in a community. Other people see it. They have opinions. Managing those opinions prevents conflicts.

Keep it neat. Messy installations with cables everywhere and crooked panels irritate people. Clean, professional installations don't.

Choose attractive equipment. All-black panels look better than traditional blue panels. Integrated mounting systems look better than DIY contraptions.

Be proactive with communication. Tell neighbors what you're doing before they notice. "Hey, I'm installing some solar panels on my balcony this weekend. Just wanted to give you a heads up."

Address concerns quickly. If someone complains about glare, noise, or appearance, respond promptly and respectfully. Small adjustments can resolve most issues.

Comply with reasonable requests. If management asks you to adjust positioning slightly for aesthetic reasons, be flexible if possible.

Show benefits, not just features. When people ask about your installation, emphasize the money you're saving and the environmental benefits. Most people respond positively to these outcomes.

Follow rules strictly. Don't give anyone legitimate reasons to complain. Stay within regulations. Don't block pathways. Don't create safety hazards.

Most conflicts arise from surprise, perceived inconsiderateness, or actual rule violations. Avoid these and you avoid most problems.

Your Installation Gameplan

Urban balcony solar installation requires planning, attention to detail, and realistic expectations. But it's not rocket science. Thousands of people have installed successfully. You can too.

Accept your balcony's limitations. Work with what you have. Optimize where possible. Accept imperfection where necessary.

Choose mounting systems appropriate for your situation. Prioritize portability if you're renting. Prioritize stability if you're in a windy location.

Install carefully following manufacturer instructions. Document everything. Check your work.

And then enjoy the benefits of generating your own electricity, even in a less-than-perfect urban environment.

Next up: battery storage. Because generating power during the day is great, but storing it for evening use is even better.

Chapter 6: Battery Storage for Apartments

Rachel generates plenty of solar power. Her south-facing balcony in Austin produces 2-3 kWh on most days. The problem? She's at work when her panels generate that power.

Her refrigerator and modem use some of it during the day. But most of the electricity her panels produce just feeds back into the grid—giving her zero benefit because her utility doesn't offer net metering for small systems.

Meanwhile, Rachel comes home at 6 PM and starts using electricity for cooking, lights, TV, air conditioning. All from the grid. All at full price. Her solar panels sit idle in the evening darkness.

She was saving maybe $15 monthly with solar alone. Then she added a 2 kWh battery storage system for $800.

Now her panels charge the battery during the day. The battery powers her apartment in the evening. Her solar savings jumped to $45 monthly. The battery pays for itself in under two years and amplifies her solar investment dramatically.

This is why battery storage matters for apartments. The timing mismatch between generation and consumption is even more pronounced for people who work outside the home. Storage solves that mismatch.

The Evening Use Problem

Let's talk about typical apartment dweller energy consumption patterns.

6 AM - 8 AM: Wake up, shower, coffee, breakfast, getting ready. Moderate electricity use. Some solar generation starting, but often minimal this early.

8 AM - 6 PM: At work or school. Minimal home electricity consumption. Just refrigerator, modem, maybe some phantom loads. Meanwhile, solar panels are generating maximum power.

6 PM - 11 PM: Home. Cooking, lights, TV, computers, AC or heat, phone charging, all appliances. Maximum electricity consumption. Zero solar generation.

11 PM - 6 AM: Sleeping. Low consumption except for refrigerator, AC/heat. Zero solar generation.

See the problem? **Peak solar generation and peak consumption barely overlap.**

Without storage, your solar panels generate electricity when you don't need it, and you buy electricity when you do need it. You save money only on daytime baseline consumption—that refrigerator and modem.

With storage, your panels charge the battery during the day. The battery discharges in the evening when you need power. Your self-consumption rate (the percentage of solar power you actually use) jumps from maybe 30% to 70-90%.

Higher self-consumption means dramatically better economics.

Battery Sizing That Makes Sense

Balcony solar systems generate 1-3 kWh daily. How much storage do you need?

The math is straightforward:

If your panels generate 2 kWh per day and you're only home in evenings, you want to store most of that power. A 2 kWh battery captures one day's generation.

But you also need to think about your evening consumption. If you use 8 kWh in the evening (common for cooking, AC, entertainment), a 2 kWh battery covers 25% of your evening load. Better than nothing, but not transformative.

Battery size options and typical uses:

1.6-2 kWh: Stores one day's generation from a 400-600 watt system. Covers baseline evening loads like lights, TV, phone charging. Runs a refrigerator for 5-8 hours. Entry-level option.

3-4 kWh: Stores generation from larger systems or multiple days from smaller systems. Covers moderate evening consumption including cooking, lights, entertainment. Runs refrigerator plus several other appliances.

5-6 kWh: Substantial storage for apartments. Can cover most evening consumption except heavy AC use. Provides meaningful backup power during outages. Premium option.

7-8 kWh: Maximum practical size for balcony solar systems under 800 watts. Stores multiple days of generation. Covers nearly all evening consumption. Provides extended backup power.

For most apartment dwellers with 400-600 watt solar systems, **2-3 kWh of battery storage hits the sweet spot**. It's enough to capture most daily solar generation and cover meaningful evening consumption without being prohibitively expensive.

Lithium-ion vs LiFePO4 Battery Chemistry

Two main battery chemistries dominate the market: standard lithium-ion (Li-ion) and lithium iron phosphate (LiFePO4, sometimes called LFP).

Standard Lithium-Ion Batteries

Chemistry: Lithium cobalt oxide or lithium nickel manganese cobalt oxide

Pros:

- Higher energy density (more storage in smaller size)
- Lighter weight
- Well-established technology
- Initially less expensive

Cons:

- More fire risk if damaged or defective
- Shorter lifespan (typically 2,000-3,000 cycles)
- Performance degrades faster with age
- More temperature-sensitive

Best for: Applications where size and weight matter more than longevity. Where replacement every 5-7 years is acceptable.

LiFePO4 Batteries

Chemistry: Lithium iron phosphate

Pros:

- Much safer (extremely low fire risk)
- Longer lifespan (6,000-10,000 cycles)
- More stable across wider temperature range
- Better performance retention over time
- Can fully discharge without damage

Cons:

- Lower energy density (larger/heavier for same capacity)
- Higher initial cost (though better long-term value)

- Slightly lower voltage (not usually a problem)

Best for: Apartment installations where safety is paramount, space isn't severely limited, and long-term value matters more than upfront cost.

For apartments, LiFePO4 is generally the better choice. The safety advantages matter when you're storing batteries indoors. The longer lifespan (10-15 years vs. 5-7 years) makes LiFePO4 cheaper over the battery's life despite higher initial costs.

Modular Battery Systems

One of the best features of modern apartment battery systems is modularity. You can start small and expand later.

How modular systems work:

You buy a base unit with a battery pack and inverter. Later, you can add additional battery packs that connect to the base unit, increasing total capacity.

For example:

- Start with 2 kWh base unit ($600-900)
- Add 2 kWh expansion pack later ($400-600)
- Total: 4 kWh system built in stages

Advantages of modular approach:

Lower initial investment. Start with smaller capacity and expand when finances allow.

Match growth to needs. See how much storage you actually use before committing to more.

Spread costs over time. Pay for expansion packs as you save money from your initial solar+storage system.

Easier transport. Smaller units are easier to carry up to apartments and move when you relocate.

Flexibility. Take what you need to a new apartment. Leave excess capacity behind or sell it if your next place has different requirements.

Popular modular systems:

EcoFlow Stream series allows stacking up to four battery packs. Start with one, add up to three more.

Jackery HomePower 2000 can expand with up to three additional batteries, bringing total capacity to 8 kWh.

Anker SOLIX balcony systems offer modular expansion in 1.6 kWh increments.

Most manufacturers are moving toward modular designs because the rental market demands flexibility.

Smart Battery Management

Modern battery systems aren't dumb boxes that store and discharge power. They're intelligent devices with sophisticated management systems.

Key smart features:

AI-powered optimization learns your consumption patterns and solar generation patterns. The system predicts when to charge, when to discharge, and how much capacity to reserve. Over time, it gets better at maximizing your savings.

Weather forecasting integration checks weather forecasts and adjusts charging strategies. If tomorrow will be cloudy, it might preserve more battery capacity today. If tomorrow will be sunny, it might discharge more aggressively tonight.

Grid price tracking (where applicable) monitors time-of-use electricity rates. It charges when electricity is cheap and discharges when electricity is expensive, maximizing savings.

App-based control lets you manually override automatic settings. You can prioritize storage, prioritize immediate use, or set custom schedules.

Energy flow visualization shows real-time data: solar generation, battery charge/discharge, grid consumption, appliance usage. You can see exactly where electricity is coming from and going to.

Remote monitoring lets you check your system from anywhere. Some systems send alerts if problems are detected.

This intelligence transforms batteries from passive storage devices into active energy management systems.

Self-Consumption Modes

Most battery systems offer different operating modes:

Storage Priority Mode

Solar power charges the battery first. Only after the battery is full does excess power feed to your home. This maximizes stored energy for evening use.

Best for: People who are away during the day and want maximum evening power. Most apartment dwellers.

Consumption Priority Mode

Solar power goes to home consumption first. Only excess power charges the battery. This maximizes immediate self-consumption.

Best for: People who work from home and use electricity throughout the day. Less common for apartments.

Backup Priority Mode

Battery stays at high charge level (80-100%) for backup power during outages. Solar charges battery and powers home simultaneously.

Best for: Areas with unreliable grid power. People who prioritize backup over daily savings.

Time-Based Control Mode

Custom scheduling. Charge during specific hours, discharge during others. Fully customizable based on your patterns and preferences.

Best for: Advanced users who want maximum control. People with time-of-use electricity rates.

Balanced Mode

Automatically balances storage, consumption, and backup needs based on patterns and predictions. The system decides optimal strategy.

Best for: Most users. Let the AI figure it out.

For apartment dwellers, Storage Priority or Balanced Mode typically delivers best results.

Backup Power Reality Check

Let's be honest about what 300-800 watt solar systems with battery storage can do during power outages.

They can:

- Keep your refrigerator running for 6-24 hours (depending on battery size)
- Power LED lights throughout your apartment

- Charge phones, laptops, and small devices
- Run WiFi router and modem
- Power small fans or heaters (briefly)
- Keep essential medical devices running

They cannot:

- Run central AC or heat for extended periods
- Power electric stoves or ovens
- Run electric water heaters
- Operate washer/dryer
- Run high-wattage appliances simultaneously
- Provide whole-apartment power indefinitely

A 2 kWh battery running a 200-watt load (refrigerator) will last about 10 hours. Running 500 watts of loads (refrigerator, lights, fans, chargers)? About 4 hours.

This is **emergency backup for essentials, not whole-home power**. But for keeping food from spoiling and staying connected during outages, it's valuable.

Some battery systems have "backup mode" that keeps the battery charged specifically for outages. Others prioritize daily savings and provide backup as a secondary benefit. Choose based on your priorities.

Indoor vs Outdoor Battery Placement

Balcony solar systems generate power outdoors. But where should you store that power?

Outdoor Placement

Pros:

- Keeps battery closer to solar panels (shorter cable runs)
- Frees up indoor space

- No concerns about indoor temperature control

Cons:

- Weather exposure (rain, temperature extremes)
- Theft/vandalism risk
- Requires weatherproof enclosure
- Potential HOA/landlord objections

Requirements if placing outdoors:

- IP65 or higher weather resistance rating
- Secure mounting to prevent theft
- Temperature-rated for your climate extremes
- Protected from direct sun exposure

Indoor Placement

Pros:

- Protected from weather and theft
- More stable temperature environment
- Easier access for monitoring/maintenance
- More socially acceptable (less visible)

Cons:

- Takes up indoor space
- Requires longer cable runs from balcony
- Indoor fire safety concerns (though minimal with LiFePO4)
- May need ventilation in hot climates

Requirements if placing indoors:

- Adequate ventilation
- Away from heat sources
- Stable surface (batteries are heavy)
- Accessible for cable connections

- Fire-safe location (not blocking exits)

For apartments, indoor placement is typically better. Most apartment dwellers have closet space or corners where a battery unit can sit. Indoor placement eliminates weather concerns and theft risk while maintaining easy access.

If you must place outdoors, invest in a high-quality weatherproof battery system specifically rated for outdoor use. EcoFlow Stream Ultra X and similar products are designed for outdoor installation.

Temperature Considerations

Battery performance varies with temperature. Understanding these limits helps you place and operate your system correctly.

Operating Temperature Ranges

Standard lithium-ion: Usually 0°C to 40°C (32°F to 104°F)

LiFePO4: Usually -20°C to 55°C (-4°F to 131°F)

Outside these ranges, batteries may:

- Refuse to charge/discharge (safety protection)
- Experience reduced capacity
- Suffer permanent damage
- In extreme cases, fail catastrophically

Cold Weather Challenges

Below freezing, lithium batteries don't charge well. Some systems include heating elements that warm the battery to chargeable temperature using grid power. This works but adds cost.

If your balcony experiences sub-freezing temperatures, either:

- Choose a battery rated for cold operation

- Place the battery indoors where it stays warm
- Accept reduced performance in winter

Hot Weather Challenges

Above 104°F, standard batteries overheat and shut down. LiFePO4 batteries handle heat better but still have limits.

If your balcony gets extreme summer heat, consider:

- Indoor placement in climate-controlled space
- Shaded outdoor placement with good ventilation
- Battery systems with active cooling (fans)

Optimal Range

All batteries perform best at 20-25°C (68-77°F). This is typical indoor temperature. Another reason indoor placement often makes sense for apartments.

Product Recommendations

Let's look at specific battery systems well-suited for apartment balcony solar. These are real products available in 2025, with realistic assessments.

EcoFlow Stream Series ($1,000-2,000)

Capacity: 1.92-11.52 kWh (modular expansion)

Chemistry: LiFePO4

Features: AI optimization, app control, weatherproof (IP65), grid backup

Pros: Highly modular, excellent app, good warranty (15 years), designed specifically for balcony solar

Cons: Premium price, bulkier than some competitors

Best for: Users wanting maximum flexibility and premium features. Those who prioritize long-term quality.

Jackery HomePower 2000 Ultra ($1,500-2,500)

Capacity: 2-8 kWh (modular)

Chemistry: LiFePO4

Features: 20ms backup switching, IP65 rated, -20°C operation, 10-year warranty

Pros: Excellent cold weather performance, fast backup switching, very robust

Cons: Expensive, heavy, limited availability in some regions

Best for: Cold climates, users prioritizing backup power, those wanting maximum durability

Anker SOLIX Balcony Systems ($800-1,800)

Capacity: 1.6-6.4 kWh (modular)

Chemistry: LiFePO4

Features: App control, automatic optimization, compact design

Pros: Good value, reliable brand, integrated design, reasonable pricing

Cons: Less expansion capacity than some competitors, basic features

Best for: Budget-conscious buyers wanting quality. Those preferring simpler systems without excessive features.

BigBlue POWAFREE ($700-1,500)

Capacity: 2.5 kWh base (expandable)

Chemistry: LiFePO4

Features: Smart plugs, customizable modes, outdoor-rated

Pros: Cost-effective, good performance, smart plug feature useful for apartments

Cons: Lesser-known brand, limited service network

Best for: Value seekers, users comfortable with less mainstream brands, those wanting smart plug integration

MANLY Battery Systems ($600-1,400)

Capacity: 1.6-7.68 kWh

Chemistry: LiFePO4

Features: Modular design, wide temperature range, long cycle life (6,000+ cycles)

Pros: Excellent price-to-performance, very reliable, good warranty

Cons: Basic aesthetics, less advanced app features

Best for: Performance-focused buyers, those prioritizing battery longevity, budget-conscious shoppers

All of these systems work well for apartment balcony solar. Choose based on your budget, desired capacity, and feature priorities.

ROI Analysis for Battery Storage

Adding battery storage increases your upfront costs. Does it make financial sense?

The math depends on several factors:

Without battery:

- 400W system generates 1.5 kWh/day average
- You self-consume 30% (0.45 kWh/day)
- At $0.20/kWh, you save $33/year
- Payback on $600 system: 18 years (terrible ROI)

With 2 kWh battery:

- Same 400W system generates 1.5 kWh/day
- You self-consume 90% (1.35 kWh/day) using battery
- At $0.20/kWh, you save $99/year
- Payback on $600 solar + $800 battery = $1,400: 14 years (still not great)

Wait, that doesn't look much better. What gives?

Here's the thing: **Battery storage makes the most sense when one or more of these conditions apply:**

You have time-of-use electricity rates. If evening electricity costs $0.35/kWh while daytime costs $0.15/kWh, storing cheap daytime solar to use during expensive evening hours makes the economics much better.

You have a larger solar system. With 800W generating 3 kWh/day, there's more excess power to store and bigger savings potential.

You live in high-electricity-cost areas. In California, Hawaii, or New England where rates exceed $0.25-0.30/kWh, storage economics improve significantly.

You value backup power. If avoiding food spoilage during outages is worth money to you, factor that into ROI.

You can't use solar power without storage. If your utility doesn't offer net metering and won't credit excess generation, storage is the only way to benefit from solar.

Better ROI scenario:

- 600W system generates 2.2 kWh/day
- Time-of-use rates: $0.35/kWh evening, $0.15/kWh daytime
- Without storage: save $0.15 × 0.5 kWh/day = $27/year
- With 3 kWh storage: save $0.35 × 1.8 kWh/day = $230/year
- Battery costs $1,000
- Battery payback: 4.3 years (much better)

Battery storage isn't universally necessary. It's most valuable for:

- People with high evening consumption
- Areas with time-of-use rates
- Larger solar systems
- Unreliable grid areas
- Those who can't benefit from net metering

If you have a small system, use significant power during the day, and have favorable net metering policies, you might not need battery storage.

Safety Certifications That Matter

Battery fires are rare but devastating. Proper certifications indicate the manufacturer takes safety seriously.

Key certifications to look for:

UL 9540 - Energy storage system safety standard. Covers the complete system including battery, inverter, and controls. This is the gold standard for stationary energy storage.

UL 1973 - Battery safety standard. Covers the battery cells and pack design specifically.

IEC 62619 - International standard for lithium battery safety. Used in Europe and increasingly recognized worldwide.

IP Rating (IP65, IP67, etc.) - Indicates water and dust resistance. IP65 means dust-tight and protected against water jets. Essential for outdoor installations.

CE Marking - European compliance marking indicating the product meets safety, health, and environmental protection standards.

FCC Certification - Required for products with wireless communication features (monitoring apps, etc.).

UN 38.3 - Transportation safety certification for lithium batteries. Indicates the battery pack can be safely shipped.

Before buying any battery system, verify it has appropriate safety certifications. If a manufacturer can't provide certification documentation, that's a red flag.

LiFePO4 chemistry batteries with proper certifications are extremely safe for apartment use. Millions are in use worldwide with excellent safety records.

What Works for Apartments

Battery storage transforms balcony solar from a marginal savings tool into a meaningful investment. But it's not required for everyone.

Add battery storage if:

- You're away during peak solar generation hours
- Your utility doesn't offer fair net metering
- You have time-of-use electricity rates
- You want backup power for outages

- You have a larger solar system (600W+)

Skip battery storage (for now) if:

- You work from home and use power during solar generation
- Your utility offers good net metering
- You have a very small system (under 300W)
- Budget is extremely tight
- You're testing balcony solar before full commitment

Battery technology keeps improving. Prices keep dropping. If storage doesn't make sense now, revisit in 1-2 years. The economics might shift.

The next chapter tackles the financial side comprehensively. We've talked about technology, regulations, installation, and storage. Now let's calculate exactly how much money you'll save and how fast your investment pays back.

Because understanding the technology is interesting. But ROI is what determines if you actually install or just keep thinking about it.

Chapter 7: Financial Analysis and ROI Calculators

Kevin spent two hours building a spreadsheet. Two hours calculating exactly how much his balcony solar system would save him in Chicago.

He factored in his electricity rate ($0.14 per kWh). He estimated his system would generate 1.8 kWh daily on average. He calculated 70% self-consumption without battery storage. He included the federal tax credit. He projected 20 years of operation.

His conclusion? The system would pay for itself in 5.8 years. After that, pure savings for another 14+ years. Total 20-year savings: about $4,200 on a $700 investment after tax credits.

Kevin installed the next week.

But his neighbor Steve looked at the same numbers and decided against it. Steve's apartment had worse sun exposure (more shading), so his generation estimate was only 1.2 kWh daily. His payback period stretched to 8.7 years. Still positive, but Steve wasn't sure he'd stay in Chicago that long. He decided to wait.

Same city. Same electricity rates. Different circumstances. Different decisions.

This is why financial analysis matters. The numbers don't lie, but they're different for everyone. Your location, your sun exposure, your consumption patterns, your electricity rates—all of this affects whether balcony solar makes financial sense.

This chapter gives you the tools to run your own numbers and make informed decisions.

High Electricity Rate Cities

Some cities have such high electricity costs that balcony solar becomes a no-brainer investment. The numbers just work.

San Francisco ($0.26-0.32 per kWh)

Maria lives in the Sunset District. Her electricity rate averages $0.29 per kWh after all fees and tiers. She installed a 600-watt system for $900 (before tax credit).

Her south-facing balcony gets good sun despite San Francisco's fog. She generates about 2.1 kWh daily on average. With 60% self-consumption (she works from home part-time), she uses about 1.26 kWh of solar power daily.

Annual savings: 1.26 kWh × 365 days × $0.29 = $133

After the 30% federal tax credit, her system cost $630. Payback period: 4.7 years. Over 20 years, she saves about $2,030 after accounting for system cost.

But Maria added a 2 kWh battery for $850. Now her self-consumption is 90%. She uses 1.9 kWh of solar power daily.

New annual savings: 1.9 kWh × 365 × $0.29 = $202

Total system cost after tax credit: $1,225. Payback: 6.1 years. 20-year savings: $2,815.

Boston ($0.23-0.28 per kWh)

James pays about $0.25 per kWh in Boston. He installed a 400-watt system for $650 plus a small 1.6 kWh battery for $700.

His east-facing balcony generates about 1.4 kWh daily (less than optimal, but workable). Battery storage gives him 80% self-consumption: 1.12 kWh daily.

Annual savings: 1.12 × 365 × $0.25 = $102

System cost after tax credit: $945. Payback: 9.3 years. Not as good as Maria's situation, but still acceptable for a long-term resident.

New York City ($0.22-0.27 per kWh)

NYC electricity rates vary by borough and building type. Elena in Queens pays about $0.24 per kWh. Her 500-watt system cost $750. She skipped battery storage because her building's net metering policy is relatively fair.

Generation: 1.6 kWh daily. Self-consumption: 50% (she's away during peak sun). Usable power: 0.8 kWh daily.

Annual savings: 0.8 × 365 × $0.24 = $70

System cost after tax credit: $525. Payback: 7.5 years.

The pattern for high-rate cities:

Payback periods typically range from 3-7 years depending on system size, sun exposure, and self-consumption rates. These are strong ROI markets for balcony solar. If you live in California, Hawaii, New England, or other high-rate regions, the financial case is solid.

Medium Electricity Rate Markets

These cities have moderate electricity costs. Balcony solar still works, but requires more careful analysis.

Chicago ($0.13-0.16 per kWh)

Average rate: $0.14 per kWh. A 400-watt system generating 1.5 kWh daily with 65% self-consumption saves about $50 annually.

System cost: $600 after tax credit. Payback: 12 years.

But bump that to a 600-watt system ($850 after tax credit) generating 2.2 kWh daily with 70% self-consumption, and annual savings reach $79. Payback drops to 10.8 years.

Add a $800 battery (cost after credit: $1,410 total) to boost self-consumption to 85%, and annual savings hit $96. Payback: 14.7 years. Battery barely makes sense here without time-of-use rates.

Seattle ($0.11-0.14 per kWh)

Seattle has relatively cheap electricity but also less sun. Average rate: $0.12 per kWh.

A 500-watt system might generate only 1.3 kWh daily (cloudy climate). With 60% self-consumption: 0.78 kWh daily.

Annual savings: 0.78 × 365 × $0.12 = $34

System cost: $700 after tax credit. Payback: 20.6 years. Barely breaks even over the panel's lifespan.

For Seattle, bigger systems make more sense. A 800-watt system generating 2.0 kWh daily saves $53 annually, paying back in 18 years on a $950 system. Still marginal.

Denver ($0.12-0.15 per kWh)

Denver has cheap electricity but excellent sun. Average rate: $0.13 per kWh.

A 600-watt system generates about 2.5 kWh daily (high altitude, clear skies). With 65% self-consumption: 1.625 kWh daily.

Annual savings: 1.625 × 365 × $0.13 = $77

System cost: $850 after tax credit. Payback: 11 years. Acceptable for homeowners, questionable for renters who might move.

The pattern for medium-rate cities:

Payback periods of 10-15 years are common. Balcony solar works, but it's a longer-term investment. System sizing and sun exposure matter more in these markets. Bigger systems that maximize generation make more financial sense than small systems.

Lower Electricity Rate Areas

Cities with cheap electricity pose challenges for balcony solar economics. The technology still works, but the financial case weakens.

Houston ($0.11-0.13 per kWh)

Texas has deregulated electricity markets and relatively low rates. Average: $0.12 per kWh.

Houston gets excellent sun. A 600-watt system generates about 2.4 kWh daily. But at low electricity rates, savings are modest.

With 70% self-consumption (1.68 kWh daily):

Annual savings: 1.68 × 365 × $0.12 = $74

System cost: $850 after tax credit. Payback: 11.5 years.

For Houston residents, balcony solar is more about environmental values and energy independence than financial returns. The investment pays back, but slowly.

Phoenix ($0.11-0.14 per kWh)

Phoenix has the best solar resource in the country but relatively cheap electricity. Average rate: $0.12 per kWh.

A 500-watt system generates an exceptional 2.6 kWh daily (intense sun, minimal clouds). With 65% self-consumption: 1.69 kWh daily.

Annual savings: $1.69 \times 365 \times \$0.12 = \$74$

System cost: $700 after tax credit. Payback: 9.5 years. Better than Houston due to exceptional sun, but still long.

Phoenix has another factor: time-of-use rates. Many utilities charge more during peak afternoon hours when AC demand peaks. If you have these rates and add battery storage to shift solar generation to peak hours, economics improve dramatically.

The pattern for low-rate cities:

Payback periods of 9-15 years are typical. For financially-motivated buyers, this is marginal. But for environmentally-motivated buyers who also want savings, it works. The key is excellent sun exposure and, ideally, time-of-use rates that reward battery storage.

Climate and Sun Hours

Electricity rates tell half the story. Solar generation depends on available sunlight, which varies dramatically by region.

Annual sun hours by region:

Southwest (Phoenix, Las Vegas, Albuquerque): 5.5-6.5 peak sun hours daily. Excellent solar resource. Systems generate 30-50% more than national average.

Southeast (Miami, Atlanta, Houston): 4.5-5.5 peak sun hours daily. Good solar resource. Hot, humid summers with some afternoon clouds.

West Coast (San Francisco, Los Angeles, San Diego): 4.5-5.5 peak sun hours daily. Good solar resource. Coastal fog reduces hours in some areas.

Mid-Atlantic (New York, Philadelphia, DC): 4.0-4.5 peak sun hours daily. Moderate solar resource. Four distinct seasons affect generation.

Midwest (Chicago, Minneapolis, St. Louis): 3.5-4.5 peak sun hours daily. Moderate solar resource. Long winters reduce annual production.

New England (Boston, Portland): 3.5-4.0 peak sun hours daily. Fair solar resource. Snowy winters significantly reduce winter generation.

Pacific Northwest (Seattle, Portland): 3.0-4.0 peak sun hours daily. Fair solar resource. Cloudy climate limits generation, especially winter.

What this means practically:

A 500-watt system in Phoenix generates about 2.6 kWh daily. The same system in Seattle generates about 1.3 kWh daily—exactly half. If Phoenix electricity costs $0.12/kWh and Seattle electricity costs $0.12/kWh, they have identical payback periods despite wildly different climates.

But if Seattle electricity costs $0.24/kWh (double Phoenix), Seattle breaks even with half the sun.

The interplay of sun hours and electricity rates determines ROI. High sun + high rates = excellent ROI. Low sun + low rates = poor ROI. High sun + low rates or low sun + high rates = moderate ROI.

Breaking Down System Costs

Let's detail exactly what you're buying and what it costs.

Entry-Level System ($400-600 before tax credit)

Components:

- One 400W solar panel: $180-250
- Microinverter (400W): $120-180
- Mounting hardware (clamp-style): $40-80
- Cables and connectors: $30-50
- Monitoring (basic or none): $0-40

Total: $400-600

After 30% tax credit: $280-420

This is the minimum viable system. One panel. Basic mounting. Minimal monitoring. Best for testing balcony solar before bigger commitments or for apartments with very limited space.

Mid-Range System ($800-1,200 before tax credit)

Components:

- Two 400W solar panels: $360-500
- Microinverter(s): $220-320
- Mounting hardware: $70-120
- Cables and connectors: $50-80
- Monitoring system (Wi-Fi): $60-100
- Installation accessories: $40-80

Total: $800-1,200

After 30% tax credit: $560-840

This is the sweet spot for most apartments. Two panels generating meaningful power. Good monitoring. Quality mounting hardware. Enough capacity to offset 10-20% of consumption.

Premium System ($1,500-3,000 before tax credit)

Components:

- Two to four 400W panels: $700-1,400

- Microinverters: $400-700
- Battery storage (2-3 kWh): $600-1,200
- Premium mounting system: $120-200
- Advanced monitoring: $100-150
- Smart plugs and automation: $80-150
- Professional installation (optional): $200-400

Total: $1,500-3,000

After 30% tax credit: $1,050-2,100

This is the full solution. Maximum practical solar capacity for balconies. Battery storage for evening use. Professional monitoring and smart home integration. Either DIY installation or professional help.

DIY vs Professional Installation

Most balcony solar systems are designed for DIY installation. But sometimes professional help makes sense.

DIY Installation

Cost: $0 (just your time)

Time investment: 2-6 hours for first-time installers

Pros:

- Zero additional cost
- Learn how your system works
- Satisfaction of self-installation
- Flexibility to install when convenient
- No scheduling coordination

Cons:

- Learning curve (though not steep)

- Potential for mistakes
- No professional warranty on installation
- Takes longer than professional work

Professional Installation

Cost: $200-500 depending on complexity

Time investment: 1-3 hours (mostly watching)

Pros:

- Fast, expert installation
- Professional warranty
- Code compliance assured
- Less worry about mistakes
- Someone to ask questions

Cons:

- Significant added cost
- Scheduling required
- Loss of DIY learning experience
- May not be necessary for simple systems

When professional installation makes sense:

- Your system requires electrical panel modifications
- You're uncomfortable with any electrical work
- Your building requires licensed installation
- You want installation warranty protection
- Your mounting situation is complex (high floors, unusual balcony)
- You're installing premium system with battery storage

For standard plug-and-play systems under 600 watts with clamp mounting, DIY installation is straightforward and saves hundreds of dollars. For complex installations or if you simply prefer professional work, the extra cost may be worthwhile.

Federal Tax Credit Details

The **30% federal solar tax credit** is called the Investment Tax Credit (ITC). It's been repeatedly extended but faces elimination proposals.

How it works:

You purchase and install a qualifying solar system. When filing your federal taxes, you claim 30% of the system cost as a tax credit. This reduces your tax liability dollar-for-dollar.

Example:

System cost: $1,000 Tax credit: $300 Out-of-pocket: $700

If you owe $2,000 in federal taxes before the credit, you'll owe $1,700 after the credit. It's not a refund—it's a reduction in taxes owed.

Qualification requirements:

- System must be installed at your primary or secondary residence
- System must be new (not used)
- You must own the equipment (not leasing)
- System must meet safety standards

The renter question:

Can renters claim the ITC? Yes, if the system is permanently installed at your residence. "Permanent" doesn't mean you can't remove it when you move—it means it's installed for ongoing use, not portable use like a camping solar panel.

Clamp-mounted balcony solar that you intend to use at your residence qualifies. Purely portable systems you could take camping don't qualify.

The urgency factor:

Congress has repeatedly proposed ending or reducing the solar ITC. As of 2025, it's scheduled to continue at 30%, but legislative proposals could eliminate it. If you're considering balcony solar, installing before potential elimination makes financial sense.

A $1,000 system costs $700 with the credit, $1,000 without. That $300 difference is significant.

State and Local Incentives

Beyond federal credits, many states and localities offer additional solar incentives.

State rebate programs:

California: Some utilities offer rebates for distributed solar. Check your specific utility. SGIP (Self-Generation Incentive Program) sometimes covers small systems with storage.

Massachusetts: SMART program offers payments for solar generation. Eligibility for balcony systems varies by utility.

New York: NY-Sun initiative provides incentives. Some programs specifically target renters.

Colorado: Xcel Energy offers solar rebates in some areas.

Maryland: Solar grant programs for residents, though primarily focused on larger systems.

Local utility incentives:

Some municipal utilities and electric cooperatives offer rebates or credits for small solar installations. These vary dramatically by location.

Sales tax exemptions:

Many states exempt solar equipment from sales tax. This saves 5-10% on equipment costs in states with exemptions.

The research requirement:

Incentives change constantly. Before purchasing, research:

- Your state's clean energy programs
- Your utility's distributed solar programs
- Local municipal solar incentives
- Sales tax exemptions in your state

Database of State Incentives for Renewables & Efficiency (DSIRE) catalogs these programs. Search by your location for current offerings.

Utility Rate Structures

Not all electricity pricing is simple. Rate structures affect your savings.

Flat-Rate Billing

You pay the same price per kWh regardless of when you use electricity or how much you use (within reason).

ROI impact: Straightforward calculation. Each kWh of solar generation saves the same amount regardless of time of day.

Tiered Pricing

First X kWh per month cost $Y. Next Z kWh cost more. Higher tiers cost even more.

ROI impact: If your solar generation keeps you out of higher tiers, savings exceed the base rate. But if you're already in low tiers, savings match the low rate.

Example: Your utility charges $0.12 for the first 500 kWh monthly, $0.18 for 501-800 kWh. If you typically use 550 kWh, your solar system reducing consumption to 485 kWh saves 65 kWh at the higher $0.18 rate—better than flat $0.12.

Time-of-Use (TOU) Pricing

Electricity costs more during peak hours (typically 4-9 PM) and less during off-peak hours (overnight, early morning).

ROI impact: This is where battery storage shines. Without storage, your solar generates during lower-rate midday hours. With storage, you use solar during higher-rate evening hours.

Example: Off-peak $0.10/kWh, peak $0.28/kWh. Solar with battery storage saves at peak rates, dramatically improving ROI.

Demand Charges

Some utilities charge based on your peak monthly demand (highest usage in any 15-minute interval). Rare for residential but exists.

ROI impact: Minimal for most balcony solar systems, which don't significantly reduce peak demand unless carefully managed with battery storage.

Net Metering

Your utility credits you for excess solar generation fed back to the grid. Credit amount and terms vary widely.

ROI impact: If your utility offers generous net metering (full retail rate credit), battery storage is less necessary. If net metering is unfavorable or nonexistent, battery storage becomes critical for capturing solar value.

Check your specific utility's rate structure. Call and ask, or review your bill carefully. The structure dramatically affects ROI calculations.

Realistic Monthly Savings

Let's ground these numbers in typical scenarios.

Scenario 1: Small system, no battery, moderate rates

System: 300W generating 1.0 kWh daily Self-consumption: 50% (0.5 kWh/day) Electricity rate: $0.15/kWh Monthly savings: 0.5 kWh × 30 days × $0.15 = $2.25

Yes, $2.25 monthly. This is why very small systems often don't make financial sense unless electricity rates are extremely high. Annual savings of $27 takes forever to recover even a $400 investment.

Scenario 2: Medium system, no battery, moderate rates

System: 600W generating 2.2 kWh daily Self-consumption: 60% (1.32 kWh/day) Electricity rate: $0.15/kWh Monthly savings: 1.32 × 30 × $0.15 = $5.94

About $6 monthly, $71 annually. On an $850 system after tax credit, payback is 12 years. Marginal but acceptable for stable long-term residents.

Scenario 3: Medium system, with battery, moderate rates

System: 600W generating 2.2 kWh daily Battery: 2 kWh Self-consumption: 85% (1.87 kWh/day) Electricity rate: \$0.15/kWh Monthly savings: 1.87 × 30 × \$0.15 = \$8.42

About \$8.50 monthly, \$101 annually. On a \$1,500 system after tax credit (including battery), payback is 14.9 years. Battery barely pays for itself at moderate rates without TOU pricing.

Scenario 4: Large system, with battery, high rates, TOU pricing

System: 800W generating 3.0 kWh daily Battery: 3 kWh Self-consumption: 90% (2.7 kWh/day), used during peak hours Peak electricity rate: \$0.32/kWh Monthly savings: 2.7 × 30 × \$0.32 = \$25.92

About \$26 monthly, \$311 annually. On a \$2,100 system after tax credit, payback is 6.8 years. Excellent ROI in high-rate TOU markets.

These scenarios show the importance of system sizing, self-consumption optimization, and rate structures. The same equipment performs very differently depending on circumstances.

The Portability Value Equation

Here's something most ROI calculations miss: portability has value.

Traditional rooftop solar is a 20-25 year commitment to a single location. Move, and you abandon your investment. This makes sense for homeowners. It's terrible for renters.

Balcony solar moves with you. That portability changes the value equation.

Example:

You install a \$700 system (after tax credit) in Boston. After 3 years, you've saved \$375. Then you move to San Francisco.

Traditional thinking: You lost money. Payback was supposed to be 7 years, but you only got 3 years of savings.

Portability thinking: You take your system to San Francisco where electricity costs $0.29/kWh instead of Boston's $0.23/kWh. Your annual savings jump from $125 to $202. Your remaining payback period on the $325 remaining investment is just 1.6 years.

Your system paid for itself in 4.6 years total across two cities, not the 7 years it would have taken staying in Boston.

The relocation ROI boost:

Moving to higher-rate cities accelerates payback. Moving to sunnier climates increases generation. Many young professionals move from lower-cost cities to higher-cost coastal cities for career opportunities. Taking your balcony solar system means your investment appreciates.

This is unprecedented in home energy improvements. Every other upgrade (new windows, insulation, HVAC) stays with the property. Balcony solar stays with you.

Factor this portability into your decision. If you might move to a high-rate or high-sun location within the system's lifetime, your expected ROI improves significantly.

Financing Options

Most people pay cash for balcony solar because systems are relatively affordable ($400-2,000 after tax credit). But financing exists.

Credit card financing:

Pros: Immediate installation. Earn rewards points. Cons: High interest rates if you don't pay off quickly. Best for: People with 0% intro APR offers who can pay off during the promotional period.

Personal loans:

Pros: Fixed rates, structured repayment, builds credit. Cons: Interest costs reduce ROI. Application process required. Best for: Larger systems ($1,500+) where monthly payments fit budget but upfront payment doesn't.

Home improvement loans:

Pros: Lower rates than personal loans. Longer terms available. Cons: May require home equity (not available to renters). Application process. Best for: Homeowners installing multiple systems or premium equipment.

Manufacturer financing:

Some manufacturers offer payment plans. Terms vary but often include 12-24 month interest-free periods.

Pros: No credit check sometimes. Convenient. Promotional rates. Cons: Limited to specific brands. Higher overall prices sometimes.

The ROI impact of financing:

If you finance at 8% APR over 5 years, you'll pay about 20% total interest on the loan. This significantly reduces ROI.

Example: $1,000 system after tax credit. Annual savings: $100.

- Cash purchase: 10-year payback, $1,000 profit over 20 years.
- 8% APR 5-year loan: 11.7-year payback (including interest), $770 profit over 20 years.

Interest eats 23% of your profit. For small balcony solar systems, cash purchase usually makes more sense unless you access 0% financing.

Resale Value Considerations

When you move, you have options for your balcony solar system.

Take it with you:

Most common choice. Uninstall, pack up, reinstall at your next apartment. Cost: 3-4 hours of your time. Value: Continue using your system and earning savings.

Sell to the next tenant:

If the next tenant wants solar, offer to sell your system. Price it at 50-70% of replacement cost depending on age and condition.

Example: Your 2-year-old system cost $1,000 new (before tax credit). Offer it to the next tenant for $500-600. They save money versus buying new. You recoup part of your investment.

Sell on secondary markets:

Facebook Marketplace, Craigslist, eBay, and solar-focused forums have markets for used equipment. Expect 40-60% of new equipment cost depending on age.

Leave it behind:

If the next tenant isn't interested and you don't want to bother moving it, some people leave systems. Treat it as a sunk cost. This is wasteful but occasionally happens.

Resale value depreciation:

- Year 1-2: 70-80% of new value
- Year 3-5: 50-70% of new value
- Year 6-10: 30-50% of new value
- Year 10+: 20-30% of new value

Solar panels last 25+ years, but used equipment buyers want significant discounts. Factor this into ROI if you think you might sell within 5 years.

Most balcony solar users keep their systems indefinitely, moving them from apartment to apartment. The portability means you rarely need to sell unless upgrading to a larger system.

Environmental ROI

Not everything is about money. Environmental impact matters to many buyers.

A typical 400-watt balcony solar system generating 1.5 kWh daily prevents approximately **559 kilograms of CO2 emissions annually**. Over 20 years, that's 11,180 kg (11.18 metric tons).

To put that in perspective:

- Equivalent to taking a car off the road for about 13 months
- Equivalent to planting 180 trees and letting them grow for 10 years
- Equivalent to avoiding 1,300 gallons of gasoline consumption
- Equivalent to 30 round-trip flights between New York and Boston

An 800-watt system doubles these numbers.

If you value environmental impact, there's an intangible ROI beyond financial returns. Some people install balcony solar even when financial payback is 15+ years because they value the environmental contribution.

Additionally, as electricity grids get cleaner (more renewables, less coal), the CO2 avoidance per kWh will decrease. Installing solar now, when grids are dirtier, has higher environmental impact per kWh than installing in 10 years.

This isn't a financial argument. It's a values argument. But values matter.

Your Numbers, Your Decision

Run your own calculations using this framework:

1. **Determine your electricity rate** (check your bill; include all fees)
2. **Estimate your solar generation** (based on system size, location, sun exposure)
3. **Calculate self-consumption percentage** (with or without battery)
4. **Multiply generation × self-consumption × electricity rate** = annual savings
5. **Calculate system cost after federal tax credit** (and any state incentives)
6. **Divide cost by annual savings** = payback period in years

If payback is under 10 years, the investment makes financial sense for most people. If payback is 10-15 years, it's marginal—depends on your plans and priorities. If payback exceeds 15 years, financial ROI is weak (though environmental reasons might still motivate you).

The next chapter covers actual product selection and installation. Because understanding the financial analysis is important. But choosing the right equipment and getting it installed is what actually generates those savings.

Chapter 8: Product Selection and Installation Guide

Melissa spent a week researching balcony solar systems. She opened seventeen browser tabs. She created a spreadsheet comparing specifications. She read reviews on five different websites.

Then she got paralyzed by choices.

Should she buy the 400W system with the better inverter or the 500W system with the worse inverter? Was the $50 price difference for the upgraded monitoring system worth it? Did she need the premium mounting hardware or would basic clamps work?

Finally, her friend who'd already installed balcony solar gave her simple advice: "Stop researching. Pick a reputable brand in your budget with good reviews. Install it. You're overthinking this."

Melissa bought an Anker SOLIX 600W system the next day. It arrived. She installed it in three hours. It worked perfectly. She's been generating solar power for eight months now.

The lesson? Perfect is the enemy of done. Yes, you should research. Yes, you should compare options. But at some point, you need to make a decision and act.

This chapter helps you make that decision efficiently.

Entry-Level Systems

Budget: $400-800 (before tax credit)

This category includes single-panel systems (200-400W) designed for testing balcony solar or for apartments with severe space constraints.

Representative Products:

Bright Saver 220W System ($450-550)

Specifications:

- One 220W panel
- California-compliant microinverter with backfeed prevention
- Basic clamp mounting
- Monitoring via smartphone app
- Plug-and-play installation

Pros: Explicitly designed for US regulations. Non-profit pricing (lower margins). California-approved.

Cons: Very low wattage limits savings. Limited availability (primarily California).

EcoFlow PowerStream 400W ($600-800)

Specifications:

- Two 200W flexible panels
- 400W microinverter
- Portable mounting options
- Advanced app monitoring
- Weather-resistant design

Pros: Flexible panels work on unusual surfaces. Good app. Established brand.

Cons: Higher price per watt. Flexible panels less efficient than rigid panels.

Generic 400W Kits (Amazon/eBay, $400-600)

Various manufacturers sell complete kits with 400W panels, inverters, and mounting hardware.

Pros: Lowest cost option. Adequate for testing.

Cons: Variable quality. Limited support. Certification uncertain.

Who should buy entry-level:

- First-time solar users wanting to test before bigger investment
- Apartments with very limited space (room for only one panel)
- Budget-constrained buyers
- People unsure how long they'll stay in current apartment

Who should skip entry-level:

- Buyers with adequate budget for mid-range systems (better value per watt)
- Those with good sun exposure and space for two+ panels
- Anyone seeking meaningful electricity bill reduction (very small systems save very little)

Mid-Range Systems

Budget: $800-1,500 (before tax credit)

This is the sweet spot. Two-panel systems (600-800W total) with quality inverters, monitoring, and mounting hardware.

Representative Products:

Anker SOLIX Balcony Solar 820W ($900-1,100)

Specifications:

- Two 410W panels (all-black, 23% efficiency)
- 600W/800W switchable microinverter
- Balcony rail mounting brackets
- Anker app monitoring

- 25-year panel warranty, 15-year inverter warranty

Pros: High efficiency panels. Excellent warranty. Reputable brand. Good monitoring app. All-black aesthetics.

Cons: Premium pricing. May require import from Europe (limited US distribution).

Robinsun Performance 800 Kit ($1,000-1,300)

Specifications:

- Two 400W panels
- 800W microinverter
- Multiple mounting options (rail, wall, ground)
- Wi-Fi monitoring
- Schuko plug (European standard)

Pros: Complete kit with everything needed. Multiple mounting options included. Good European track record.

Cons: European-focused (adapters needed for US outlets). Limited US customer service.

DIY Mid-Range Build ($800-1,200)

Components purchased separately:

- Two 400W panels: $360-500 (Renogy, Newpowa, or similar)
- Two 400W microinverters: $240-360 (Enphase IQ8M or Hoymiles)
- Mounting hardware: $80-120
- Cables and connectors: $60-100
- Monitoring: $60-120 (optional)

Pros: Customizable. Can select best components for your needs. Often best value.

Cons: Requires more research. Need to ensure component compatibility. No single warranty.

Who should buy mid-range:

- Most balcony solar buyers (this is the standard choice)
- Those with space for two panels
- Buyers wanting meaningful savings (10-20% electricity offset)
- Anyone planning to stay in apartment 5+ years

Who should skip mid-range:

- Those with very limited budgets (entry-level works)
- Buyers wanting maximum performance with battery storage (premium systems better)
- Anyone with space for only one panel

Premium Systems

Budget: $1,500-3,000 (before tax credit)

Complete solutions with multiple panels, battery storage, advanced monitoring, and premium components.

Representative Products:

EcoFlow Stream Ultra System ($1,800-2,500)

Specifications:

- Two to four 400W panels
- Stream microinverter with AI optimization
- 1.92-11.52 kWh LiFePO4 battery (modular)
- IP65 weatherproof rating
- Advanced app with energy forecasting
- Smart plugs for load management
- 15-year warranty

Pros: Complete ecosystem. Modular battery expansion. Excellent app. Weather-resistant. AI optimization.

Cons: Premium price. Bulky battery. Requires dedicated space.

Jackery HomePower 2000 Ultra ($2,000-3,000)

Specifications:

- Two 400W panels
- 2-8 kWh battery (expandable)
- 20ms backup switching
- -20°C to 55°C operation
- IP65 rated
- App monitoring
- 10-year warranty

Pros: Excellent cold-weather performance. Fast backup switching. Very robust. Proven reliability.

Cons: Expensive. Heavy (60+ lbs). Premium features many won't use.

Custom Premium Build ($1,600-2,800)

Components:

- Three to four 400W panels: $540-800
- Multiple microinverters: $360-640
- 2-3 kWh battery: $700-1,200
- Premium mounting: $150-250
- Advanced monitoring and smart home integration: $150-250

Pros: Fully customized. Can select exact components. Maximum performance potential.

Cons: Complex to design and install. Requires significant research. Compatibility challenges.

Who should buy premium:

- Serious solar enthusiasts wanting maximum performance
- Those with high electricity consumption and costs
- Buyers planning 10+ year stays or willing to move system multiple times
- Anyone prioritizing backup power during outages
- Those with time-of-use rates benefiting from battery storage

Who should skip premium:

- Budget-conscious buyers (mid-range delivers better value)
- First-time solar users (start smaller and expand)
- Those unsure about long-term commitment
- Renters who might face landlord restrictions on larger systems

Trusted Manufacturers

Brand reputation matters. Quality manufacturers stand behind their products. Unreliable manufacturers disappear when you need warranty support.

Tier 1: Premium Established Brands

EcoFlow: Founded 2017. Known for portable power stations. Expanded into balcony solar. Excellent engineering. Strong customer support. Premium pricing.

Jackery: Founded 2012. Leader in portable power. HomePower series designed for balcony solar. Proven reliability. Good warranty support.

Anker (SOLIX line): Founded 2011. Consumer electronics giant. SOLIX balcony solar launched 2022. Benefits from Anker's distribution and support network.

Tier 2: Solid European Brands

Robinsun: Spanish company focused on plug-and-play systems. Strong European presence. Limited US distribution but growing.

Sunflower Solar: French balcony solar specialist. Good products. European standards. Adapters needed for US.

Tier 3: Component Manufacturers

Renogy: Solar panels and components. Good quality-to-price ratio. DIY-friendly. Strong Amazon presence.

Enphase: Microinverter leader. Industry-standard reliability. Used in professional installations and DIY builds.

Hoymiles: Microinverter manufacturer. Good alternative to Enphase. Lower cost. Adequate quality.

What to avoid:

- No-name brands with zero online presence
- Companies with no US-based support
- Manufacturers with consistent poor reviews
- Systems with no safety certifications
- Prices significantly below market (usually quality corners cut)

Stick with established manufacturers. The $50-100 you might save buying sketchy equipment isn't worth the risk of failure, safety issues, or no warranty support.

US Market Availability (2025 Update)

The US balcony solar market is developing rapidly. Here's what's actually available as of 2025:

Widely Available:

- Individual solar panels (400W) from multiple manufacturers via Amazon, specialized solar retailers
- Microinverters from Enphase and Hoymiles
- Mounting hardware from various sources
- Basic complete kits from smaller brands

Limited Availability:

- Premium European systems (Anker SOLIX, Robinsun) often require ordering from European websites with international shipping
- Battery-integrated systems designed for balconies (growing but limited)
- Systems explicitly certified for US plug-and-play use

Not Yet Available:

- Mass-market options in big-box stores (Home Depot, Lowe's)
- Many European brands that haven't entered US market
- True turn-key solutions with guaranteed US regulatory compliance in all states

The DIY advantage:

Because complete systems are limited, many US buyers assemble systems from components. This requires more research but offers:

- Better availability (components are easier to source than complete systems)
- Customization to your specific needs
- Often better value
- Ability to source best-in-class individual components

The market is expanding quickly. Expect major retailers to stock balcony solar within 1-2 years as regulations become clearer and demand increases.

Reading Solar Panel Datasheets

Panel specifications matter. Here's how to read technical specs:

Power Rating (Wp - Watts peak):

Example: 400Wp

This is maximum power output under ideal test conditions (bright sun, cool temperature, specific angle). Real-world output is typically 75-85% of rated power.

Higher is better, but also usually bigger/heavier.

Efficiency (%):

Example: 21.5%

Percentage of sunlight converted to electricity. Standard panels: 18-20%. High-efficiency: 21-23%. Premium: 24%+.

Higher efficiency means more power from smaller panel size. Valuable for space-constrained balconies.

Dimensions:

Example: 1722 mm × 1134 mm

Physical size. Check that panels fit your balcony space. Standard residential panels are roughly 5.5-6 feet tall by 3-3.5 feet wide.

Weight:

Example: 21 kg (46 lbs)

Important for mounting hardware load calculations and installation difficulty. Heavier doesn't necessarily mean better.

Voltage Ratings:

Voc (Open circuit voltage): Maximum voltage with no load. Example: 37.4V

Vmp (Voltage at maximum power): Optimal operating voltage. Example: 31.2V

Isc (Short circuit current): Maximum current. Example: 13.8A

Imp (Current at maximum power): Optimal current. Example: 12.8A

These specs need to match your inverter's input requirements. Most microinverters handle wide voltage ranges, but verify compatibility.

Temperature Coefficient:

Example: -0.35%/°C

How much power decreases as temperature rises above 25°C (77°F). Lower (less negative) is better.

Hot sunny days reduce panel output due to temperature effects. Premium panels have better temperature coefficients.

Warranty:

Product warranty: 10-15 years typical (covers defects)

Performance warranty: 25-30 years typical (guarantees minimum output)

Standard performance warranty: 90% output at 10 years, 80% at 25 years.

Inverter Selection

The inverter converts DC from panels to AC for your home. Critical component.

Wattage Rating:

Match inverter capacity to total panel wattage. For two 400W panels, use 600W or 800W inverter. Slight undersizing is acceptable (panels rarely hit maximum output simultaneously).

Efficiency:

Good microinverters: 95-97% efficiency. Premium: 97%+. Higher efficiency means less power lost in conversion.

Input Voltage Range:

Inverter must handle your panel's voltage. Most microinverters accept 16-60V, covering most panels.

Grid Standards:

US inverters must meet UL 1741 and IEEE 1547 standards for grid connection. Verify compliance.

Features:

Rapid shutdown: Required for safety. All modern inverters include this.

Monitoring: Some inverters include WiFi monitoring. Others require separate monitoring devices.

Multiple MPPT trackers: Premium feature allowing independent optimization of multiple panels. Useful for partial shading.

Recommended Options:

Enphase IQ8M: 330W, 97% efficiency, excellent reliability, grid-forming capability, $180-220

Hoymiles HM-600: 600W, 96% efficiency, good value, WiFi monitoring, $130-170

APsystems QS1: 300W per panel (four-panel capacity), 97% efficiency, integrated monitoring, $250-320

Monitoring Systems

Knowing how much power you're generating is half the fun. Monitoring systems provide real-time and historical data.

Built-In Inverter Monitoring:

Many modern microinverters include WiFi and connect to smartphone apps. Shows:

- Current power generation
- Daily/monthly/yearly production
- Panel-level data (if multiple panels)
- Historical trends
- Alerts for problems

Pros: Included with inverter purchase. Seamless integration. Reliable.

Cons: Limited to inverter manufacturer's app features.

Third-Party Monitoring:

Systems like Sense or Emporia monitor your entire home's electricity, including solar generation.

Pros: Detailed whole-home energy data. Identify what appliances use power. Better analytics.

Cons: $150-300 additional cost. More complex installation.

Simple Power Meters:

Plug-in power meters ($20-50) on the outlet show real-time generation. No historical data or apps.

Pros: Cheap. Immediate feedback. Easy setup.

Cons: No data logging. Must look at device for information. Basic.

What Most People Choose:

Start with whatever monitoring your inverter includes. If you want more detail later, add third-party monitoring.

Real-time monitoring is motivating. Watching power generation on sunny days provides satisfaction beyond just financial savings.

Installation Tools You Actually Need

Most balcony solar installation requires minimal tools.

Essential:

- Adjustable wrench or socket set (for tightening mounting hardware)
- Screwdriver set (Phillips and flathead)
- Measuring tape
- Level (optional but helpful)
- Voltage tester/multimeter (verify outlet function)

Optional but Useful:

- Drill with concrete bits (only if mounting requires drilling)
- Cable ties and organizers
- Rubber padding (if not included with mounts)
- Electrical tape

- Weatherproof sealant (for cable protection)
- Safety glasses

Total cost for tools: $40-100 if you don't already own them. Most people have basic tools already.

What You Don't Need:

- Specialized solar tools
- Advanced electrical equipment
- Expensive specialty items

Plug-and-play systems are designed for installation with basic household tools.

Safety Testing Before First Use

Before plugging in your system, verify everything is correct.

Visual Inspection:

- Panels are securely mounted
- No cracks or visible damage to panels
- All electrical connections are tight
- No exposed wiring
- Cables are organized and protected
- Mounting hardware is snug (not loose, not over-tightened)

Outlet Testing:

Use outlet tester ($10) or multimeter to verify:

- Outlet is properly grounded
- Voltage is correct (120V in US)
- No wiring faults

Connection Testing:

Before plugging in inverter:

- Verify panel connections are secure
- Check that MC4 connectors are fully clicked
- Ensure inverter is properly attached to mounting system

First Power-On:

- Plug in inverter
- Check for inverter status lights (consult manual for normal status)
- Verify monitoring system connects (if applicable)
- Watch for any unusual sounds, smells, or visible issues
- On sunny day, confirm panels are generating power

If anything seems wrong, unplug immediately and diagnose the issue.

Troubleshooting Common Issues

Problem: No Power Generation

Possible causes:

- Not enough sunlight (check on sunny day)
- Inverter not powered on
- Loose connections
- Inverter shutdown due to grid issues

Solutions: Check connections. Verify outlet works. Wait for sunny conditions. Consult inverter manual for error codes.

Problem: Lower Than Expected Generation

Possible causes:

- Shading (partial or full)
- Dirty panels

- Panel angle suboptimal
- Inverter malfunction
- Unrealistic expectations

Solutions: Clean panels. Adjust angle if possible. Verify no new shading sources. Check monitoring data over multiple weeks (daily variation is normal).

Problem: Inverter Won't Connect to Monitoring

Possible causes:

- WiFi signal too weak on balcony
- Router settings blocking device
- App configuration incorrect

Solutions: Check WiFi signal strength. Reset inverter WiFi. Reconfigure app connection. Consult manufacturer support.

Problem: Mounting Hardware Feels Loose

Possible causes:

- Insufficient tightening during installation
- Hardware loosening over time (vibration, wind)
- Wrong size clamps for railing

Solutions: Retighten hardware. Add rubber padding for better grip. Check weather conditions (high winds may require stronger mounting).

Problem: Lower Generation in Winter

Possible causes:

- Shorter days (normal)
- Lower sun angle (normal)
- Snow covering panels

- More frequent cloudy days

Solutions: This is expected seasonal variation. Clean snow off panels. Accept that winter generation is lower. Annual production is what matters.

Maintenance Requirements

Good news: Balcony solar needs minimal maintenance.

Every 3-6 Months:

- Clean panels (water and soft cloth)
- Check that mounting hardware is tight
- Verify no new shading sources (growing trees, new buildings)
- Inspect cables for wear or damage

Annually:

- Deep clean panels if needed
- Review monitoring data to ensure performance hasn't degraded
- Check warranty status (important if approaching warranty expiration)

As Needed:

- Remove snow after winter storms
- Clean panels after dusty/sandy weather
- Check connections if performance drops suddenly

What maintenance is NOT needed:

- Inverter maintenance (no user-serviceable parts)
- Panel maintenance beyond cleaning
- Electrical system testing (unless problems arise)

Total maintenance time: 1-2 hours annually. Balcony solar is remarkably low-maintenance.

When and How to Upgrade

You might want to upgrade your system after initial installation.

Adding Panels:

If you start with one panel and want more power, adding panels is straightforward. Just get compatible panels and inverter capacity to handle them.

Adding Battery Storage:

Most mid-range systems can add battery storage later. Check compatibility between your inverter and available battery systems.

Upgrading Inverter:

If your inverter fails or you want better features, replacing it is simple. Panels remain. Just swap inverter.

Upgrading Monitoring:

Can add or upgrade monitoring systems any time. Doesn't affect solar generation.

When NOT to Upgrade:

- If current system meets your needs (don't upgrade just to upgrade)
- If you're moving soon (wait until new location)
- If upgrade cost doesn't justify additional savings

Moving with Your System

One of balcony solar's best features: it moves with you.

Dismounting:

Reverse installation process. Takes 1-3 hours depending on system complexity.

Packing:

- Original packaging is ideal (save boxes if possible)
- Bubble wrap panels carefully (fragile)
- Coil cables without sharp bends
- Place mounting hardware in labeled bags
- Protect inverter during transport

Transport:

- Panels should lay flat during transport
- Avoid stacking heavy items on panels
- Secure everything so nothing slides during transport

Reinstallation:

Follow original installation process. Faster second time because you've done it before.

Moving Checklist:

1. Document your current installation (photos)
2. Test system one final time before dismounting
3. Dismount carefully
4. Pack securely
5. Inspect new balcony before reinstalling
6. Reinstall at new location
7. Test system at new location

Warranty Considerations

Warranty coverage protects your investment. Understand what's covered.

Panel Warranties:

Product warranty: Covers manufacturing defects. Typically 10-15 years.

Performance warranty: Guarantees minimum power output. Typically 25-30 years with degradation schedule (90% at 10 years, 80% at 25 years).

Inverter Warranties:

Typically 10-15 years for quality microinverters. Some extend to 20-25 years.

Battery Warranties:

Usually 5-15 years depending on chemistry. LiFePO4 typically longer than standard lithium-ion.

What Warranties Cover:

- Manufacturing defects
- Early failure
- Below-spec performance

What Warranties Don'T Cover:

- Damage from improper installation
- Physical damage (dropping panels, impact damage)
- Weather damage beyond rated specifications
- Unauthorized modifications
- Normal wear and tear beyond warranty terms

Claiming Warranties:

Keep all receipts and documentation. Register products with manufacturers. Document any issues with photos/data. Contact manufacturer support promptly if problems arise.

What You Need to Do Next

You've learned about regulations, installation, financial analysis, and now product selection. Theory is done. Time for action.

Your next steps:

1. Decide your budget
2. Measure your balcony space
3. Check local regulations (Chapter 3)
4. Select a system from appropriate category
5. Purchase equipment
6. Install following safety procedures
7. Monitor and enjoy your solar power

Balcony solar isn't complicated. Choose equipment. Install it. Generate electricity. Save money.

The appendices that follow provide quick-reference resources: regulatory charts, template letters, calculators, checklists, and additional resources.

Everything you need is here. The only thing missing is your decision to act.

Appendix A: State-by-State Regulatory Quick Reference

States with Clear Plug-and-Play Legislation:

Utah - HB 429 (2025): Systems up to 1,200W exempt from interconnection requirements. No utility approval needed. Portability explicitly recognized.

States with Pending Legislation (as of late 2025):

Vermont - Proposed 800W plug-and-play exemption. Expected passage in 2026.

Maryland - Tenant rights legislation for balcony solar pending.

Pennsylvania - Micro-generation framework under development.

States with Favorable Conditions (No Specific Legislation):

California - No specific plug-and-play law, but solar access laws protect reasonable installations. Utilities vary in policy. Bright Saver operates with backfeed-prevention systems.

Colorado - Strong solar access laws. Many utilities accommodate small systems informally.

Massachusetts - SMART program may cover small systems. Check specific utility.

States with Neutral/Unclear Status:

Most states fall here. No specific prohibition, no specific permission. Gray area navigation required.

Recommended Action by State:

- Utah: Install freely under 1,200W
- California: Use backfeed-prevention systems or contact utility
- Vermont/Maryland/Pennsylvania: Wait for legislation or install carefully
- All other states: Review Chapter 3, contact utility, proceed cautiously

For Latest Updates:

Check DSIRE (Database of State Incentives for Renewables & Efficiency) and state legislature websites for current status.

Appendix B: Landlord Permission Letter Template

Template 1: Initial Request Letter

[Use this when first requesting permission]

Dear [Landlord/Property Manager Name],

I am writing to request approval for installing a portable solar panel system on my balcony at [Your Address, Unit Number].

System Overview: I plan to install a [wattage]W solar panel system consisting of [number] solar panels with clamp-style mounting brackets. The system requires no drilling, screws, or permanent modifications to the property.

Key Details:

- Total weight: approximately [weight] pounds
- Dimensions: [dimensions]
- Installation method: Removable clamps on balcony railing
- Installation time: Under 2 hours
- Removal time: Under 1 hour
- All equipment meets UL safety standards

Benefits:

- Reduces my electricity costs
- Demonstrates environmental responsibility
- Increases property appeal to sustainability-minded tenants
- Creates no maintenance burden for property management
- No structural modifications to property

Responsibilities: I will be fully responsible for:

- All installation, maintenance, and eventual removal
- Any costs associated with the system
- Ensuring the installation causes no damage to property
- Compliance with all applicable regulations
- Full restoration of balcony to original condition upon move-out

Safety and Insurance:

- My renter's insurance policy [does/will] cover the equipment
- System includes automatic safety shutdown features
- Installation complies with [city/state] building codes
- No fire or electrical hazards are created

I have attached photos of similar installations showing minimal visual impact. I would welcome the opportunity to discuss this request and address any questions or concerns.

Thank you for your consideration. I look forward to your response.

Sincerely, [Your Name] [Contact Information] [Date]

Template 2: Follow-Up Letter (After Initial Denial or No Response)

Dear [Landlord/Property Manager Name],

I am following up on my request dated [date] regarding installation of a portable solar panel system at [address].

I understand you may have concerns, and I would like to address them:

If concerned about damage: The system uses non-permanent clamps with rubber padding. No drilling or permanent attachments. I am willing to pay an additional security deposit of $[amount] to cover any potential concerns.

If concerned about appearance: I have selected an all-black panel design that is minimally visible and maintains property aesthetics. I can provide exact specifications and visual mockups.

If concerned about liability: My renter's insurance covers the equipment and I can add the property as an additional insured party at no cost to you.

If concerned about regulations: I have verified the installation complies with [city/state] regulations and I will handle all required notifications to utilities.

I would greatly appreciate the opportunity to discuss this further. Many apartment communities now accommodate balcony solar as a tenant amenity. I am a responsible tenant who has [paid rent on time/maintained the property well/been a tenant for X years] and I view this as a long-term investment in my residence here.

Please let me know if you would be willing to reconsider or if there are specific conditions under which you would approve the installation.

Thank you again for your time and consideration.

Sincerely, [Your Name] [Contact Information] [Date]

Appendix C: ROI Calculator Spreadsheet

Simple ROI Calculation Formula:

```
System Cost (after tax credit) = Equipment Cost × 0.70

Daily Solar Generation (kWh) = Panel Watts ÷ 1000 ×
Peak Sun Hours × 0.80

Daily Usable Solar = Daily Generation × Self-
Consumption %

Annual Savings = Daily Usable Solar × 365 × Electricity
Rate

Payback Period (years) = System Cost ÷ Annual Savings

20-Year Net Benefit = (Annual Savings × 20) - System
Cost
```

Example Calculation:

- Equipment Cost: $1,000
- System Cost after 30% tax credit: $700
- Panel Wattage: 600W
- Peak Sun Hours: 4.5 hours/day
- Daily Generation: 0.6 × 4.5 × 0.80 = 2.16 kWh
- Self-Consumption: 70%
- Daily Usable Solar: 2.16 × 0.70 = 1.51 kWh
- Electricity Rate: $0.18/kWh
- Annual Savings: 1.51 × 365 × $0.18 = $99.23
- Payback Period: $700 ÷ $99.23 = 7.1 years
- 20-Year Net Benefit: ($99.23 × 20) - $700 = $1,284.60

Variables to Adjust for Your Situation:

- Equipment cost (depends on system chosen)
- Peak sun hours (look up for your city)

- Self-consumption % (estimate based on your schedule)
- Electricity rate (check your utility bill)

Appendix D: Installation Safety Checklist

Pre-Installation:

☐ Balcony is structurally sound
☐ Weight capacity verified (if high floor)
☐ Weather is calm (no high winds, rain)
☐ All tools and equipment are ready
☐ Installation area is clear of obstacles
☐ Fire escape routes are not blocked
☐ Instructions have been read completely

During Installation:

☐ Wearing appropriate protective equipment
☐ Panel handling is careful (fragile)
☐ Mounting hardware is appropriate for railing type
☐ Clamps are tightened evenly (not over-tightened)
☐ Rubber padding is used to protect railing
☐ Electrical connections are secure (MC4 connectors clicked)
☐ Cables are organized and protected
☐ No cables create trip hazards
☐ Inverter is securely mounted
☐ Outlet has been tested (grounded, correct voltage)

Post-Installation Testing:

☐ Visual inspection shows secure mounting
☐ No loose connections or exposed wiring
☐ Panels are stable (gentle rocking test)
☐ Inverter powers on correctly
☐ Status lights show normal operation

☐ Monitoring system connects (if applicable)
☐ Power generation confirmed on sunny conditions
☐ No unusual sounds, smells, or heat
☐ Installation documented with photos
☐ All packaging and materials cleaned up

Ongoing Safety:

☐ Check mounting security every 3-6 months
☐ Inspect cables for wear periodically
☐ Keep panels clean for optimal performance
☐ Monitor for new shading sources
☐ Verify system continues operating normally

Appendix E: Product Comparison Matrix

Entry-Level Systems (200-400W):

Product	Wattage	Price Range	Pros	Cons	Best For
Bright Saver 220W	220W	$450-550	CA-compliant, non-profit pricing	Very low output	California residents testing solar
EcoFlow PowerStream 400W	400W	$600-800	Flexible panels, good app	Higher $/watt	Unusual mounting surfaces
Generic 400W Kit	400W	$400-600	Lowest cost	Variable quality	Budget testing

Mid-Range Systems (600-800W):

Product	Wattage	Price Range	Pros	Cons	Best For
Anker SOLIX 820W	820W	$900-1,100	High efficiency, warranty	Premium price, limited US distribution	Quality-focused buyers
Robinsun Performance 800	800W	$1,000-1,300	Complete kit, multiple mounting	European focus	Those comfortable with imports
DIY Component Build	600-800W	$800-1,200	Customizable, best value	Requires research	DIY enthusiasts

Premium Systems (with Battery Storage):

Product	Wattage	Battery	Price Range	Pros	Cons	Best For
EcoFlow Stream Ultra	800W+	1.92-11.52 kWh	$1,800-2,500	Modular, AI optimization	Premium price	Tech-savvy buyers
Jackery HomePower 2000 Ultra	800W	2-8 kWh	$2,000-3,000	Cold-weather rated, robust	Expensive, heavy	Cold climates, backup priority

Appendix F: Utility Contact Information by State

How to Contact Your Utility:

1. Find your utility name on your electricity bill
2. Search "[Utility Name] interconnection" or "[Utility Name] distributed generation"
3. Call customer service: Ask for "interconnection department" or "distributed generation department"
4. Email inquiries: Get responses in writing when possible

Major Utilities by Region:

California:

- Pacific Gas & Electric (PG&E): 1-877-743-4112
- Southern California Edison (SCE): 1-800-655-4555
- San Diego Gas & Electric (SDG&E): 1-800-411-7343

New York:

- Con Edison: 1-800-752-6633
- National Grid: 1-800-930-5003

Texas:

- Various retail electric providers (deregulated market)
- Search "Texas electric provider [your city]"

Massachusetts:

- Eversource: 1-800-592-2000
- National Grid: 1-617-469-2300

Illinois:

- ComEd: 1-800-334-7661

For complete list by state, visit: https://www.eia.gov/electricity/

Appendix G: Glossary of Technical Terms

AC (Alternating Current): Electricity that periodically reverses direction. Standard for home electrical systems (120V/60Hz in US).

Anti-Islanding: Safety feature that shuts down solar inverter when grid power fails, preventing backfeed to utility lines during repairs.

DC (Direct Current): Electricity flowing in one direction. Solar panels generate DC power.

Inverter: Device that converts DC electricity from solar panels to AC electricity for home use.

kW (Kilowatt): Unit of power. 1,000 watts. Measures instantaneous power output or consumption.

kWh (Kilowatt-hour): Unit of energy. Amount of power used over time. A 1,000W appliance running for 1 hour consumes 1 kWh.

LiFePO4 (Lithium Iron Phosphate): Battery chemistry known for safety, longevity, and temperature tolerance. Preferred for residential energy storage.

MC4 Connectors: Industry-standard waterproof connectors for solar panel wiring. Click and lock design.

Microinverter: Small inverter attached to each solar panel (or pair of panels), allowing independent panel optimization.

MPPT (Maximum Power Point Tracking): Technology that optimizes power output from solar panels by finding the voltage and current combination that produces maximum power.

Net Metering: Utility policy crediting customers for excess solar power fed back to grid.

Peak Sun Hours: Measure of solar irradiance. One peak sun hour equals 1,000 watts per square meter of sunlight. Used to estimate solar generation.

Rapid Shutdown: Safety feature requiring solar systems to quickly de-energize when disconnected or during emergencies.

Self-Consumption: Percentage of solar-generated electricity used directly by the home (vs. exported to grid).

String Inverter: Single large inverter handling multiple solar panels connected in series. Contrast with microinverters.

Time-of-Use (TOU) Rates: Electricity pricing that varies by time of day, typically higher during peak demand periods.

UL (Underwriters Laboratories): Independent safety certification organization. UL-certified equipment meets specific safety standards.

Voc (Open Circuit Voltage): Maximum voltage produced by solar panel with no load connected.

Watt (W): Unit of power. Measures rate of energy transfer.

Appendix H: Additional Resources and Community Groups

Online Communities:

Reddit:

- r/SolarDIY - DIY solar installation discussion
- r/solar - General solar topics, news, advice
- r/homeautomation - Smart home integration with solar

Facebook Groups:

- "Balcony Solar Systems - USA"
- "Apartment Renters Going Solar"
- "DIY Solar Power"

YouTube Channels:

- "Will Prowse" - DIY solar tutorials and product reviews
- "Solar Surge" - Balcony and portable solar focus
- "EnergySage" - Consumer solar education

Educational Resources:

DSIRE (Database of State Incentives for Renewables & Efficiency): https://www.dsireusa.org/ Comprehensive database of solar incentives by state

Energy Sage: https://www.energysage.com/ Solar marketplace and educational resource

SEIA (Solar Energy Industries Association): https://www.seia.org/ Industry organization with consumer resources

NREL (National Renewable Energy Laboratory): https://www.nrel.gov/ Research institution with solar data and tools

Regulatory Information:

NEC (National Electrical Code): https://www.nfpa.org/NEC Electrical safety standards (updated every 3 years)

UL Standards: https://www.ul.com/ Safety certification information

Local Building Codes: Contact your city's building department or search "[Your City] building codes solar"

Product Reviews and Comparisons:

Amazon Customer Reviews: Real user experiences with specific products

Solar Review Websites:

- SolarReviews.com
- EnergySage marketplace reviews
- YouTube product review channels

Advocacy Organizations:

Vote Solar: https://votesolar.org/ Advocates for solar-friendly policies

Environment America: https://environmentamerica.org/ Campaigns for renewable energy adoption

Local Solar Cooperatives: Many cities have local solar co-ops that provide group purchasing, education, and advocacy

Tools:

Sun Position Calculator: https://www.suncalc.org/ Shows sun path for your location

PVWatts Calculator (NREL): https://pvwatts.nrel.gov/ Estimates solar production by location

Google Project Sunroof: https://sunroof.withgoogle.com/ Analyzes your location's solar potential (limited to some areas)

Moving Forward

You now have a complete guide to balcony solar: the technology, regulations, financial analysis, product selection, installation procedures, and ongoing management.

The information is here. The market is developing. The regulations are improving. The technology is proven.

What happens next depends entirely on you.

Thousands of Americans have already installed balcony solar and are generating their own electricity. Millions more have the potential to do so. The 53 million households without rooftop solar access don't need to stay locked out of the solar revolution.

Balcony solar isn't perfect. It won't power your entire apartment. It won't make you energy-independent. It won't eliminate your electric bill.

But it will reduce your costs. It will give you some control over your energy. It will let you participate in clean energy generation. And it will do all of this in a portable, renter-friendly format that moves with you.

The barriers are falling. Utah led the way. Other states are following. The momentum is building.

The question isn't whether balcony solar will become mainstream in America. The question is whether you'll be an early adopter or wait until everyone else has already installed.

Either choice is valid. But now you have the information to make an informed decision.

The sun is shining. The technology works. The economics are improving.

Your balcony is waiting.

References

Core National Sources

- Database of State Incentives for Renewables & Efficiency (DSIRE). (2025). State incentives for renewables & efficiency. NC Clean Energy Technology Center, N.C. State University. https://www.dsireusa.org/
- National Renewable Energy Laboratory. (2025). PVWatts® calculator. U.S. Department of Energy. https://pvwatts.nrel.gov/
- U.S. Energy Information Administration. (2025). Electricity data browser. U.S. Department of Energy. https://www.eia.gov/electricity/data/browser/
- Solar Energy Industries Association. (2015). Residential consumer guide to solar power. https://www.seia.org/
- Vote Solar. (2025). Policy advocacy for solar by state. https://votesolar.org/

Industry Guides, News, and Notable Company Pages (U.S., Germany, Global)

- Bright Saver. (2025). Plug-and-play solar for U.S. homes. https://www.brightsaver.org/
- CBC News. (2025, August 30). Plug-in balcony solar panels could mean cheaper power. https://www.cbc.ca/news/climate/plug-in-balcony-solar-panels-1.7618883

- EcoFlow. (2025). STREAM series balcony solar plants. https://www.ecoflow.com/
- Energy Storage News. (2025, April 15). EcoFlow launches balcony power plant with new modular battery approach. https://www.ess-news.com/2025/04/15/ecoflow-launches-balcony-power-plant-with-new-modular-battery-approach/
- Enphase Energy. (2025). IQ series microinverters. https://enphase.com/
- GreenLancer. (2025, March 10). Solar energy for rental homes: A guide for tenants and landlords. https://www.greenlancer.com/post/solar-energy-rental-home
- Hoymiles. (2025). Microinverters for residential solar. https://www.hoymiles.com/
- IntegrateSun. (2024, November 30). Can solar panels be an option for renters? https://www.integratesun.com/post/can-solar-panels-be-an-option-for-renters
- Jackery. (2025). Battery storage for balcony solar systems. https://eu.jackery.com/
- MANLY Battery. (2025, July 7). Balcony solar and storage battery tips. https://manlybattery.com/balcony-solar-and-storage-battery/
- PV Magazine. (2025, April 28). Balcony solar gains US foothold with plug-and-play systems. https://www.pv-magazine.com/2025/04/28/balcony-solar-gains-us-foothold-with-plug-and-play-systems/
- Robinsun. (2025). Plug-in solar kits for balconies. https://robinsun.es/
- SolarReviews. (2024). Solar panels for rental homes and apartments. https://www.solarreviews.com/blog/solar-panels-for-rental-homes-and-apartments

- SolarTech Online. (2025, August 8). Solar panels for apartment balconies: Complete 2025 installation guide. https://solartechonline.com/blog/solar-panels-apartment-balcony-guide/
- Texas Solar Energy Society. (2025, September 4). Balcony solar panels: The future of urban energy? https://txses.org/balcony-solar-panels-the-future-of-urban-energy/

European and International Official Sources

- Bundesnetzagentur. (2024). Balcony power plant installations in Germany. Federal Network Agency.
- MDPI. (2025). Barriers to balcony solar and plug-in distributed energy resources in the United States. Energies, 18(8), 2132.
- SlashGear. (2024, May 5). Installing solar panels on your balcony: Is it a good idea? https://www.slashgear.com/1573861/installing-solar-panels-on-balcony/

Supplementary, Academic, and Governmental

- U.S. Department of Energy. (2024). Plug and play solar PV for American homes. Solar Energy Technologies Office.
- Utah State Legislature. (2025). HB 429: Small portable solar generation devices. Legislative records.